BIODYNAMICS FOR BEGINNERS

BIODYNAMICS FOR BEGINNERS

Principles
and Practice

Hugh J. Courtney
and other contributors

Portalbooks ≈ 2024

Portal books
An imprint of SteinerBooks
834 Main Street, PO Box 358
Spencertown, New York 12165
www.steinerbooks.org

Copyright © 2024
The Josephine Porter Institute for Applied Bio-Dynamics

All rights reserved. No part of this publication may be reproduced, stored in a retrieval system, or transmitted, in any form or by any means, electronic, mechanical, photocopying, recording, or otherwise, without the prior written permission of the publisher.

LIBRARY OF CONGRESS CONTROL NUMBER: 2023945809

ISBN: 978-1-938685-41-5

Contents

Introduction by Stewart Lundy	ix
HONORING THE COW	1
Man as Synthesis of Eagle, Lion, and Ox/Cow	3
How the Cow Supports Human and Earth Evolution	5
Substance and Forces—Physical and Spiritual	7
Spiritualizing the Earth: Humanity's Debt to the Animal Kingdom	8
On the Harvesting of the Preparation Sheaths	14
THE SCIENTIFIC BASIS OF BIODYNAMICS	16
Inspiration in Science	17
Scientific Validation of Biodynamics	20
Scientific Objectivity	26
SENSITIVE CRYSTALLIZATION: REVEALING THE LIFE FORCES BEHIND BIODYNAMICS	31
500: THE FOUNDATIONAL HORN MANURE PREPARATION	46
FURTHER THOUGHTS ON MAKING BD500	51
501: HORN SILICA PREPARATION	58
502: HOW TO MAKE THE YARROW PREPARATION	70
The Stag Bladder	72
Drying the Inflated Stag Bladder	75
Preparing to Stuff the Stag Bladder	76
Stuffing the Stag Bladder with Yarrow	76
Summer Exposure	79
Ready for Fall	79
Burying the Yarrow Preparation	80
Digging up the Yarrow Preparation	81
Storage	82
Uses	82

Achillea millefolium Esoterica	83
503: Chamomile: The Healer for People and Plants	90
Making the Chamomile Preparation at JPI	92
Harvest and Preparation of the Blossoms	92
Harvesting the Intestines	94
Stuffing the Intestines	97
Burying the Preparation	100
Storage	103
Effects of the Preparation	104
504: The Stinging Nettle Preparation	108
505: How to Make the Oak Bark Preparation	124
Harvesting and Preparing the Oak Bark	124
Obtaining and Making Ready the Skull	129
Stuffing the Brain Cavity	131
Choosing the Site to Bury the Skulls	132
Recovering the Oak Bark Preparation	133
The Oak Bark Preparation:	
Organ of Living Thinking (in the Farm Organism)	138
506: Dandelion: Messenger of Heaven	146
History of Dandelion	147
Botany	147
A Dynamic Weed	148
Dandelion as Medicine	149
Jupiter	151
Dandelion in the Kitchen	152
506: Dandelion Compost Preparation	154
507: How to Make the Valerian Preparation	166
Picking the Flowers	166
Time of Year to Harvest	167
Time of the Moon	167
Time of Day	167

Processing the Flowers	168
Extracting the Liquid	168
Fermentation Time	170
The Filtering Process	171
For Use	172

THE VALERIAN PREPARATION: ADDITIONAL PERSPECTIVES — 173

508: REVEALING THE HIDDEN FORCES OF EQUISETUM ARVENSE — 184
  The Course Structure — 187
  Using Equisetum: Two Approaches — 191
  Equisetum in the Context of the Cosmic Forces
    in Our Solar System — 197
  Practical Application of an Abstract Intellectual Theory — 200
  Equisetum as Primary Support
    for "Peppering" and for the Farm Individuality — 202

PEST "PEPPERS" AND HOMEOPATHIC DILUTION — 205
  Part I — 205
  Part II — 221

THE THREE KINGS PREPARATION — 230
  Making and Using the Preparation — 235

PREPARATIONS AS BEINGS:
  THE THREE KINGS PREPARATION OF HUGO ERBE — 240
  Method — 242
  Protocol for Hugo Erbe's Three Kings Biodynamic Preparation — 244
  The Proportions — 245
  The Amount of Prepared Substance and the Amount of Water — 245
  To Summarize — 246

EXPANDING THE BIODYNAMIC PREPARATIONS — 248

AMETHYST 501 — 261
  Cited Works — 266
  Further Reading — 270

Introduction

Stewart Lundy

> *"Though we do not wholly believe it yet, the interior life is a real life, and the intangible dreams of people have a tangible effect on the world."* – JAMES BALDWIN, Nobody Knows My Name

Describing the invisible world in terms of the visible requires a delicate touch. Since we can only see what's in front of our eyes, to begin to speak of the invisible world, we must be fluent in analogies based on what we can see and be confident that the universe is a unified whole, not a mere amalgam of unrelated pieces. As the philosopher Aristotle says, we must begin with what we know. Without a unifying world conception, there is no impetus for scientific inquiry and no basis for knowledge. Life would be unlivable if everything did not always already possess an inner kinship. Just as we cannot see the wind itself but we can see what it does, we may not be able to perceive the tones of subtle "forces" but any of us can witness their effects if we put them to use. Similarly, we cannot see for ourselves the inner workings of the soil, but we can see the effects of the secrets of the soil in how the emerging plants express themselves above ground.

To journey into the realm of the invisible—which, if we're honest, is *most* of the world—we must keep in mind the entire time that any analogy based on the sense-perceptible world will invariably be lacking. Our senses are not designed to perceive most of the cosmos but, rather, are tuned to a narrow bandwidth of a mere fraction of a percent of all possible information available. As such, any analogy based on that tiny fraction of a percent will be incomplete at best. But if we approach the world with a "soft" gaze, we can allow disparate viewpoints to become a composite image of a living whole.

There is a story of blind monks grasping at different parts of an elephant: one thinks the leg is the trunk of a tree, one imagines the tail is a rope, one thinks the ear is a large leaf, and so forth. Each similarity, by itself, is incomplete, but that does not mean they are each individually false. The individual ideas of "tree" and "rope" and "leaf" are all, by themselves, correct concepts but they are all *misapplied* to the elephant. If each blind monk were to trade places with the others successively and try to reconcile these various perspectives, a clearer image of the whole—the elephant itself—would emerge. Each limited view is a legitimate vantage point as far as it goes—after all, there is a *likeness* between the elephant's tail and a rope—but when these experiences of separate concepts are reconciled, an even greater concept of the whole emerges.

Some particularly bright stars, such as Rudolf Steiner, seem to have reached far beyond the limits of what is sense-perceptible (and therefore beyond what is externally empirical) into realms inaccessible to ordinary everyday consciousness. But if we need specialized sense organs to perceive light, there are analogous inner "organs" we require to perceive the dark light of the invisible world. If we want to glimpse the subsensible world, we might use a microscope. But if we wish to understand the meaning of the kaleidoscopic panoply of our ever-changing sense perceptions, we need to be able to intuit macrocosmic interconnections that no external technology can reveal for us. If we are like blind monks grasping at pieces, Steiner is like someone who could grasp the encompassing idea "elephant" while the rest of us are busy arguing from our smaller, one-sided perspectives.

How Steiner reached his clear-sightedness is somewhat beside the point for our current discussion of biodynamic agriculture. If I lack eyes, I cannot perceive light myself, but my blindness does not negate the empirical existence of colors for anyone else with eyesight. I may have no experiential point of reference to evaluate whether "red" or "blue" exist—or even what those terms mean—but that does not mean that colors as such have no reality merely because I personally cannot experience them. Nevertheless, Steiner did not expect blind faith in what he observed. Anything Steiner disclosed he consistently said should be *tested* and empirically validated.

Steiner never asked people to "believe" what he said. On the contrary, Steiner asked his audience to *think* what he said and also to *test* what he said. The trouble these days seems to be that many of us cannot even begin to entertain a new idea unless we already believe it. The result of closed-mindedness is the inability of various groups to find common ground—blind monks insisting that an elephant is a "rope" or a "tree." But if we dismiss out of hand a new idea merely because it seems "weird," we will never grow outside what we already think we know. And if we do not submit an idea to empirical testing and document results, we cannot expect others to believe our claims. Funding for biodynamic research is often in short supply partly because it promises to make farmers buy less fertilizer, not more. Biodynamics, as part of the cultural sphere, aims to make farms more profitable and not extract greater profits from farmers themselves. As such, biodynamic research is nurtured primarily by the initiative of magnanimous donors seeking to help farmers worldwide become more ecologically and economically sustainable.

When the insights of Steiner drawn from anthroposophy (living consciousness of one's humanity) are applied to agriculture, we find a distinctly humane approach to farming that we can call *biodynamics*. The novelty of biodynamics is not that it is sustainable. Indigenous practices worldwide—including in Europe—were long established before the industrial revolution—in many cases for millennia, as F. H. King demonstrates in *Farmers of Forty Centuries*, for instance. What makes biodynamics special is that it is a post-industrial attempt to return to natural wisdom. As Owen Barfield describes anthroposophy in his introduction to *The Case for Anthroposophy*:

What differentiates anthroposophy from its traditional predecessors, both methodologically and in its content, is precisely its post-revolutionary status. It is, if you are that way minded, the perennial philosophy; but, if so, it is that philosophy risen again, and in a form determined by its having risen again, from the psychological and spiritual eclipse of the scientific revolution.[1]

Biodynamics is distinguished by the fact that it is a *return* from and through a mechanized prodigal society, not a regression. It is a return in a *new* way, like returning to your childhood home as an adult. As T. S. Eliot sings in *The Four Quartets*: "We shall not cease from exploration/And the end of all our exploring/Will be to arrive where we started/And know the place for the first time." Anthroposophy is coming to "know the place for the first time" and arriving where we started, albeit ourselves changed. As Steiner suggested, we should continue to farm in any way that works and, to that, *add* the biodynamic preparations to sustainable practices. As such, biodynamics is universally applicable wherever sound farming practices are already employed. Above all, biodynamics is a practical application of insights gleaned from anthroposophy. One might ask whether you need to believe in anthroposophy to use biodynamics, and the answer is: of course not! Steiner did not want sycophants. Steiner said, we need "active fellow workers—no mere executive organs."[2] It is not enough for people to follow orders from a centralized authority; we need creative coworkers who can not only fulfill tasks but add more than they are asked to give.

As a man well versed in scientific literature, Steiner would likely have had a special place in his heart for those who are skeptical about the claims of biodynamics and yet dare to test it with objectivity. After all, science should not shy away from what is weird—*superstition* does that—but rather, science tests whether a hypothesis is

---

1 Owen Barfield, introduction to Steiner, *The Case for Anthroposophy*.
2 Steiner, *Agriculture Course: The Birth of the Biodynamic Method*, p. 60.

Introduction

repeatable. As the science fiction author Arthur C. Clarke famously wrote, "Any sufficiently advanced technology is indistinguishable from magic." The theoretical basis of biodynamics may be somewhat difficult to grasp, but anyone can experience its concrete results.

In his *Agriculture Course*, Steiner advances a falsifiable (testable) claim that these biodynamic preparations, when properly employed, tend to help produce healthier crops when *added* to sound sustainable farming practices. Biodynamics does not aspire to replace agriculture as we know it, nor is it opposed to innovations. Biodynamics merely seeks to restore a piece of what is often neglected, though it cannot succeed without good farming practices as a rich foundation. Despite what some might suggest, biodynamics does not need to take much time—a stir here and there amounts to a few hours in the scope of an entire year. Good farming takes a lot of time, but the addition of biodynamics is a negligible investment of time by comparison, especially given its benefits to the soil. How long do we sit in traffic throughout our brief lives? We don't stop driving because of that. If you make your own preparations on your farm, it undoubtedly adds a few more hours of work, but you're cultivating "indigenous microorganisms," now popularly referred to as IMOs, in a new proactive way by making these special preparations.

It was only with my discovery of the microbiological work of Ehrenfried Pfeiffer's *Soil Fertility, Renewal and Preservation* that I found my point of entry into biodynamics. While the preparations are irreducible to microbiology, they certainly contain a rich range of beneficial microbes. The work investigating these "smallest entities" exploded by Pfeiffer, Maria Thun, and Eugen and Lili Kolisko, among others, bore enough fruit to merit repeating and expanding upon their experiments.

If the right nutrients and their activities are absent in the soil, the plants above will be unable to express themselves to their fullest potential. Biodynamics as such starts with the soil by offering a way to help restore to the exhausted earth qualities that have been depleted by centuries of agricultural extraction. At first glance, it may look like how you already farm or garden. Biodynamics offers a way for regenerative and sustainable practices to improve synergistically. As such, biodynamics is applicable to almost infinitely varied agricultural settings. Perhaps you turn beds as usual, but now you spray out the horn manure or Pfeiffer's Field and Garden Spray immediately before cultivating.

A Zen saying tells us that you should meditate half an hour a day unless you're too busy, then meditate a full hour a day. If we don't have time to use biodynamic preparations on our farm, we probably already neglect other important things. Biodynamic soil tends to have a microbiological diversity closer to that of wild virgin land and healthy nutrient cycling. Biodynamic preparations, when made in each region, express an infallible connection to the *terroir* of that specific place. As biodynamic farmers, we are cultivating a unique ecosystem fostered by the innate biodiversity of each distinct region. We begin with the natural rhythm that already exists and finesse it to produce human food alongside rich biodiversity.

Biodynamics is a reorientation of agriculture to its primary purpose—not commerce, but life. Anthroposophy is not a fanatical movement, nor is it an iconoclastic movement. As Steiner says, we need cow horns to make the biodynamic preparations, but we do not need to be "bull-headed" about it. Only what is directly inimical to life should be rejected. What is useful and can remain fruitful for human flourishing should be retained. In a word, Steiner's suggestion of a humane economy harmonizes well with E. F. Schumacher's concept of "appropriate technology" in his idea of Buddhist economics outlined in *Small is Beautiful: Economics as if People Mattered*. Whether this integrates with Steiner's more esoteric Christian approach to problems, as proposed in his *Threefold Social Order*, one needs only to recall how Schumacher himself said it might have been called Christian economics—but then no one would have taken him seriously. Biodynamics is not about shifting a local economy to become dependent on others

Introduction

but rather about empowering individual spaces and communities to be ever more food- and energy-independent. After all, food is the primary form of human energy. Nurturing the human capacity for freedom through agriculture and nutrition is at the heart of biodynamics. Biodynamics aims not at reinventing the wheel but restoring whatever happens to be missing to otherwise sustainable farming practices. As such, biodynamics is not a complete system, nor does it particularly propose to be. Biodynamics is *added* to sound farming practices as an "appropriate technology" that can be produced within the limits of a diversified farm. The ideas of biodynamics are free to all. The spiritual world holds no patents on thoughts. You don't have to be clairvoyant to put ideas to the test.

Properly utilized, the biodynamic preparations help farmers worldwide become freer from expensive inputs. While the ingredients for the various preparations may seem obscure, these were intentionally drawn from easily sourced materials on the farms of Steiner's day. Offal and weeds—that's all the preparations are made from. And, it should be remembered, Steiner said virtually all of the biodynamic herbs could easily be replaced—with the possible exception of Stinging Nettles (*Urtica dioica*). Does this mean that Stinging Nettles *doesn't* have a suitable substitute, or is it not *easily* found? On this question, more research is merited.

Most of the biodynamic herbs are weeds that grow easily throughout the world, and ruminant offal can be found anywhere animals are processed. While the biodynamic preparations may seem odd, their ingredients were intended to be made from ready-at-hand materials within the limits of a diversified, sustainable farm. Though the reason for the combination of a particular flower with a particular animal membrane may seem obscure, they are meant to restore qualities that have been depleted and to do so without requiring outside inputs. It is almost as if the flowers and the sheaths "rhyme" in a suprasensory way, though we don't have to be able to hear their concordance to use them and see results.

The revolutionary significance of biodynamics is that farmers do not need to import nearly as many synthetic fertilizers to produce nutrient-dense food. If we were to wed biodynamics with sound farming practices, farms could become nitrogen independent, learning to inhale more of what they need directly from the atmosphere itself.

The late Alex Podolinsky's work with farmers demonstrates the power of biodynamics on large swaths of land without any synthetic fertilizers. This is done by suitable deep-ripping with a subsoiler, diverse cover cropping, rotational grazing, *plus* the biodynamic preparations. Podolinsky discusses his process extensively in his three-volume set of *Bio-Dynamic Introductory Lectures*, encouraging us to imagine humus not as a material substance but as a "living process" of many things transforming into many other things. "There is no 'permanent' humus. Humus exists only at the height of a PROCESS of continuous becoming. That is, *in status nascendi*. 'Permanent' humus would be dead material."[3] It is as if humus is not a single compound but a dynamic nexus where a collective process of ongoing transformation is sustained. Potential energy is added to the soil so that plants can express that as kinetic energy in growth forms. Restoring the "etheric" life potential to the soil is like recharging a battery. Depleted soil must be recharged, which Nature might take centuries to restore.

When we seek to re-enliven soil, we seek to restore its inner dynamism so that it is not dead dirt but a nexus where the intersection between earth and cosmos weave in an intimate and ongoing exchange. Human beings, at their best, humanize the environment. At our worst, when our cleverness serves mere consumption, we desecrate every landscape we touch. But when we use technology judiciously while remaining open to life from above, we may become "wise as serpents and innocent as doves."

---

3   Alex Podolinsky, *Living Agriculture*, p. 8.

Introduction

The torchbearers for biodynamic work are relatively few and far between. Some tend the flame for a short period, others for a lifetime. Rudolf Steiner instructed Ehrenfried Pfeiffer, who brought biodynamics to North America. Pfeiffer taught Josephine Porter, and Josephine Porter passed her mantle to Hugh Courtney (1932–2020). Hugh Courtney founded the Josephine Porter Institute in 1985. He began *Applied Biodynamics* in 1992 as a quarterly newsletter for the Josephine Porter Institute.

The following collection of articles is from *Applied Biodynamics*, which blossomed into more than a newsletter. These articles were assembled primarily by Hugh Courtney as examples of some of his favorite pieces from his time overseeing *Applied Biodynamics*. Additional pieces have been selected for the purpose of filling in gaps in practical knowledge around making all the biodynamic preparations. There are many books on biodynamics, and this book is not intended to replace any of them. Instead, these articles are meant to disclose the *practical* side of biodynamics more than the theoretical aspects. This introduction is intended to provide some familiarity with the conceptual terrain from which biodynamics grows for anyone seeking to explore the spiritual wellspring that gives rise to the practical articles of *Applied Biodynamics*.

For those seeking a broader theoretical context for biodynamics, it is almost impossible to imagine biodynamics without the writings of Goethe, specifically *The Metamorphosis of Plants*. I would venture so far as to say that Steiner is almost inconceivable without Goethe, and biodynamics is almost unthinkable without *The Metamorphosis of Plants*. There is more packed into the little *Metamorphosis* than one might initially suspect—it deserves deep contemplation. As further context for biodynamics, Steiner's early books concerning Goethe's scientific work, including *Goethe's World View* and *Nature's Open Secret,* are both dense collections of excellent thoughts. But of all the books Steiner wrote, one held a special place in his heart: *Intuitive Thinking as a Spiritual Path: A Philosophy of Freedom* (also translated as *The Philosophy of Freedom: The Basis for a Modern World Conception*). In short, *Intuitive Thinking* is not about a life of mere subjective feeling but rather about how to be a free human being, which is also at the heart of what a biodynamic farm aims to support. As a bit of forewarning to readers, if you find Steiner's writing difficult, that is normal and even, according to Steiner in his introduction to *Theosophy*, intentional. As a muscle does not develop without resistance—and as astronauts lose muscle mass in outer space—there is an inner capacity you *exercise* by wrestling with difficult passages page by page and sometimes even sentence by sentence. Small weights repeatedly lifted build larger muscles. The effort itself is already a kind of success.[4]

It is reasonable to ask why these medicinal soil remedies are called "preparations." In the first place, they have to be called something. Still, within the context of Rudolf Steiner's oeuvre, there are three stages of disciplined spiritual development as articulated in *How to Know Higher Worlds*: 1) preparation, 2) illumination, and 3) initiation. The first preparatory stage lays the foundation for receptivity to luminosity from above, providing the basis of initiation. If a room is dark because a window is obscured, you must first remove the blockage so the light can stream in and *then* you can see new creative ways to use your free initiative within the freshly illuminated space. As St. Thomas Aquinas says, "The reality of things is their light."[5]

Consider the analogy of learning to ride a bicycle. First, you might work with training wheels, striving to develop a sense of equipoise; then you practice (and fail repeatedly) trying to balance on your own until finally there is that magical eureka moment where it *clicks*. The

---

4   "This book cannot be read the way people ordinarily read books in this day and age. In some respects, its readers will have to work their way through each page and even each single sentence the hard way. This was done deliberately; it is the only way this book can become what it is intended to be for the reader. Simply reading it through is as good as not reading it at all. The spiritual scientific truths it contains must be *experienced*; that is the only way they can be of value" (Steiner, introduction to *Theosophy*).

5   Commentary to *Liber de causis* 1,6,

Introduction

preparatory training wheels give way to illumination by the new capacity, which in turn delivers an initiation into a new kind of freedom theretofore impossible.

Biodynamics, as such, provides *preparations* for the soil so that the plant and farm "organism" can be more receptive to light from above and be initiated into freedom. The preparations help root the plant into properly enlivened soil and improve photosynthesis, which in turn improves the nutrient content of fruits and vegetables and enhances the carbon sequestration activity performed by cover crops.

The ideal of any farm is to provide as much of its own fertility needs as possible from within its own resources—to be free from external economic compulsion. As such, the biodynamic preparations are just that: *preparatory*. The biodynamic preparations do not replace sensible farming practices. They are not a panacea. Someone living an irregular and unethical life but incorporating sporadic spiritual exercises cannot expect to make much progress. Similarly, someone applying the biodynamic preparations on top of unsound farming practices cannot expect good results. When negligent people use biodynamics, they may claim that biodynamic preparations "don't work" when it is the farm itself that was already not working. But if the preparations are used in combination with sensible farming, biodynamics empowers plants to be themselves and, by extension, facilitates a greater capacity for human beings to be themselves.

Steiner's destiny, by his own account, was originally to compose a book of "peasant wisdom," though he said he instead took on another man's destiny to edit Goethe's work. We should consider his words seriously because I do not think he failed to fulfill his destiny. What else is the biodynamic impulse but the fully embodied expression of this kind of peasant wisdom in a practical form? Despite its relationship to his destiny, Steiner did not seek out speaking on agriculture. He was nearly coerced to give these lectures when the son of Count Keyserlingk showed up at Steiner's door and refused to leave without a commitment from him to speak on agriculture. Steiner complements the Keyserlingk family as having an "iron will." Steiner even traveled (against his physician Dr. Ita Wegman's recommendations) to offer this humble flower out of the anthroposophic impulse, which contains the rarefied destiny he had suppressed for so long.

Steiner gave the course (published as *Agriculture*) in Koberwitz, Silesia (in what is now part of Poland), in 1924. He died the next year. One might almost say that biodynamics is Steiner's swan song.[6] The main audience of the original agriculture course was not farmers, and Steiner expected attendees to be familiar with the contents of his books, *Theosophy* and its sequel, *An Outline of Esoteric Science*. Neither are easy books, so if you find *Agriculture Course* confounding, it should be considered the third book of a demanding trilogy. Those seeking the broader context of biodynamics may risk exploring it, but for the rest of us who want to put ideas to *work* now, the following collection of articles is for you.

My relationship with biodynamics began by pestering Hugh Courtney with countless questions. While he did direct me to other resources, he primarily recommended that I attend his hands-on practicum that fall, where he would instruct a small group of gardeners, farmers, and other friends of the earth to make the biodynamic preparations. I didn't make it the first year I was corresponding with him, but I did the next year and for the last seven years of his life. Everything was new. It was hard to imagine that this could end and almost harder to imagine that it had ever begun. By the end of each day, I was exhausted in body and soul. Just trying to assimilate all the newness was overwhelming. But I returned year after year, and the process integrated as a conscious part of me. But with familiarity, even something as unusual as stuffing manure in horns or oak bark into skulls can seem ordinary, obvious even. How easily we forget the miracle of life and how unnecessary, and therefore free, it is!

---

6   For historical context on the lectures themselves, *The Agriculture Course Koberwitz, Whitsun 1924* by Peter Selg, and *The Birth of a New Agriculture* by Adalbert von Keyserlingk are both valuable resources.

Introduction

Once I grasped the external process, I understood *how* to make these unusual preparations, but I still didn't understand *why* they were made like this.

It took me many years—over a decade—to begin to have a living appreciation for the biodynamic preparations. It was not until reading Jakob Böhme's work, specifically *The Aurora*, that the "life processes" behind the scenes in biodynamics began to germinate in me. *The Aurora* is an extended, repetitive "seed meditation" for those willing to undertake the exercise. Some may recognize elements from Steiner's seed and plant meditations in *How to Know Higher Worlds*. It was through Böhme's work that Steiner's *Theosophy* finally became unlocked.[7] Before the living experience of the world in Böhme, everything else had been blind grasping, like a raccoon "washing" its food in water—I had only been feeling out the shape of the preparations with external questions grounded in the blindness of mere sensory experience. But since the preparations are drawn from suprasensory ideas, no amount of external skepticism can illuminate a single spiritual idea. Merely external questioning is akin to grasping a locked box but with a blindfold covering your eyes: with enough attentiveness to what your fingers are sensing, you can still evoke a fairly clear inner image of what you are sensing in your hands—the *gestalt* of the object. Owen Barfield calls this approach "dashboard knowledge" in *Saving the Appearances*—like a child who climbs into a vehicle and starts pushing buttons, we can get external results without knowing anything about how the inner operations of the world actually work. Even if we become proficient at driving a car, that does not mean we have the knowledge to repair the engine if something goes wrong. Look at our environment today; it's hard to say that something hasn't gone wrong, but the kind of thinking that broke the world is not

---

7 As I later discovered, Steiner himself says, "One needs only to know Paracelsus and Jacob Boehme to know theosophy. Everything that they wrote is given from a deep spring, with immense deepness and magic power" (Steiner, Berlin, May 3, 1906).

going to repair it. Superficial knowledge only goes skin deep. To this point, Gary Lachman quotes Saint-Martin in *Caretakers of the Cosmos*, saying, "The proof that we are regenerated is that we regenerate everything around us."[8] If we wish to heal the world, we have to restore invisible qualities in the soil and in our own souls.

The good news is that just because I cannot see something does *not* mean that everyone else is blind to the same thing. Thankfully, the fruits of esoteric insights from those with insights into the invisible world may be communicated with the rest of us, and they can be put into immediate practice by absolutely anyone. We do not have to be clairvoyant to be sensible practical farmers, though Steiner did suggest in *Agriculture Course* that a farmer's intimate relationship with animals and plants has an innate tendency to produce *clairsentience*—clear and objective senses, which are the prerequisite to what Goethe called "exact sensorial imagination." Steiner did not suggest that farmers have any natural tendency toward mystical clairvoyance—clear sight into spiritual worlds—but rather that farming tends to produce clear, practical objective *sensing* in any attentive farmer. If we are not objective about what we're sensing, any inner image we produce from that sensory experience will be distorted. But the attentive farmer tends to look closely and pay attention to the sights, sounds, and smells of life—because the livelihood of a farmer depends on it. There is no room for mere abstract philosophical speculation, guesswork, or expensive gambles when your livelihood depends on maintaining healthy living beings.

Unlike academia, where errors can go unchecked and yet a tenured professor retains his or her salary, a farmer's relationship to feedback from the world of practical action is direct and costly. Over time, a farmer develops a cautious, conservative temperament because the profit margin in farming is so narrow. This gives what Steiner calls the "gruff exterior" of the farmer, yet inwardly they're inclined toward a

---

8 Lachman, *Caretakers of the Cosmos*, p. 104.

Introduction

deep tolerance for others. "If it works for you, go for it!" tends to be the seasoned farmer's attitude toward his neighbors.

Hugh made preparations for longer than my lifespan. During that time, he learned to evaluate when a preparation was good or bad, with often little more than a glance. This objective sensory capacity approached the *clairsentience* that Steiner suggests would arise in daily practical work with living things. This doesn't mean that Hugh's conclusions drawn from his clear-eyed senses were necessarily always right. Steiner himself said that a normal, intelligent human being may correct a clairvoyant, and Hugh would never have claimed special insight into the spiritual world. Though not a farmer, Hugh Courtney learned to make quality biodynamic preparations from practical experience.

One of the first things that told me that there was something to the biodynamic preparations wasn't their effectiveness on the soil–though that also proved itself valuable–but rather my own attempts to make them. I failed to produce anything but a smelly mess: this impressed upon me that there were indeed quality standards. And if there were quality standards, there was something valuable at work here, something more to strive toward. If my first attempts at making biodynamic preparations had succeeded easily with no need for careful attention, I would have had reason to doubt the value of biodynamics altogether. Because I made bad preparations, I knew there must also be good preparations and went out in search of them.

Hugh Courtney's favorite preparations were the oak bark preparation and the horsetail (*Equisetum arvense*) fermented decoction. For the oak bark preparation, he insisted on using fresh heads to remain as much as possible "within the realm of the living," as Steiner recommends in *Agriculture Course*.

During one of my last visits to Hugh's farm, he conscripted me to drive his old tractor down the steep muddy hill in Woolwine, Virginia, to his burial site for the oak bark skulls. Now, I live on flatland and did not want to drive this machine with bad brakes in the rain on slippery clay. I had no experience driving a heavy tractor on steep inclines, but Hugh insisted because no one else there was familiar with riding a tractor–or no one else spoke up, that is. Hugh walked in the evening of his life with a cane, but he took a brisk pace down the hill in front of the tractor. As we entered a wooded area, the tractor tires began to slip on mud and leaves. Hugh never turned back to see if the tractor was about to barrel into him. The man walked like someone who knew he couldn't die until it was time to die. I, on the other hand, was tense and poised to turn the wheel and crash into the nearest tree instead of running over Hugh. Those minutes on that slope feel much longer, in my recollection, but we made it to the bottom safely.

One of the last things Hugh Courtney told me during my final visit was, "The next generation needs to take over biodynamics now." This was in the fall of 2019, barely half a year before he crossed the threshold. It was at this final workshop that he handed over the oak bark preparation and asked me to teach the attendees. For a number of years, I resisted taking up the responsibility of helping produce biodynamic preparations. Hugh had warned me against the complexities of working with the Josephine Porter Institute and even more ominously that "biodynamics accelerates karma." But it was already too late.

Hugh liked to remind attendees of his workshops that, "Biodynamics is *not* an intellectual path. Biodynamics is a path of the *will*." As such, working from the outside inward, biodynamics begins with practical work, which gradually transforms inner feeling, which then transmutes thought. From that transfigured way of thinking, new feelings arise, and new actions are conceived. This is the obverse to many other initiatic approaches, even within anthroposophy, which has a tendency to begin with the head and hopes that having the right thoughts leads to the right actions. However, as we are embodied beings—sometimes it makes a greater difference to take practical steps first and let the inner transformation follow.

Introduction

Biodynamics begins with the hands, moves to the heart, and culminates with the head. I've said this many times before, but the greatest effect the biodynamic preparations have is not on the farm, though that is not insignificant, but rather *on the farmer*. Combining things that, by chance, would never be united, transforms the imagination. The practice of making the biodynamic preparations with your own hands creates new pathways and thereby opens up new possibilities for free expression. It's not a tangible change at first, but, before long, new possibilities begin to disclose themselves that previously were never conceived. Before long, you, yourself, are transfigured.

I had been asked to join the board of the Josephine Porter Institute but declined. Later, I accepted a small media manager role working with *Applied Biodynamics*, but I am writing now as the creative director of the Josephine Porter Institute as of 2023, whose role is expanding the reach of biodynamics, working with our board of directors, assisting our farm manager with maintaining and improving biodynamic preparation quality, coordinating research, and facilitating the actualization of the biodynamic ideal of the self-sufficient organism at the farm at the Josephine Porter Institute. Hugh saw the purpose of biodynamics as fundamentally spiritual and oriented toward human freedom; and this is the reason we undertake the task of healing the earth.

The following set of articles from *Applied Biodynamics* introduces readers to some of the basics of making the biodynamic preparations. These articles were compiled primarily by Hugh Courtney himself as examples of some of his favorite articles during his time overseeing *Applied Biodynamics*. However, some additional articles were selected to supply fullness to this collection. What is discussed in these articles is not the only way to make preparations successfully, but it is, for the most part, Hugh's way of doing so. Some people have undoubtedly made more preparations, while others have made preparations longer. Still, there is no one I know of in North America who has made as many preparations *and* for as long a time as Hugh

Courtney. Though Hugh had many students, and his light passed to many other candles, The Josephine Porter Institute for Applied Biodynamics continues to nurture a special heart of this sacred flame born out of practical anthroposophy and love for the world.

Even toward the end of his life here, if you asked Hugh a question, he would likely give a non-answer like, "Let me know what you find out." As frustrating as that can be for a neophyte seeking ready-made answers, it deflected people's focus away from him and back to the work at hand. I suspect Hugh often knew more answers than he shared. But as muscles only grow with resistance, Hugh did not shy away from making you wrestle with biodynamics by having you test it yourself. As such, I will direct you to the work remaining ahead of you.

It is my earnest prayer that the following articles provide the reader with sufficient technical know-how to begin to learn how to make quality preparations yourself and to remember we do this *so that the earth may be healed.*

Honoring the Cow

Hugh J. Courtney
Applied Biodynamics, *no. 55, 2006*

Each year when we are preparing for the harvest of the several bovine sheaths used to make the biodynamic preparations, we choose to observe the solemnity of the occasion by leading the participants in what has become a ritual tradition here at the Josephine Porter Institute We all gather the night before the slaughter and read several passages from one of Rudolf Steiner's most profound descriptions of the meaning of the animal kingdom in human and "Earth" evolution. After sharing his words, we end with a reading of Christian Morgenstern's poem "Washing of the Feet." A visit is then made to the barnyard where the chosen animal is then silently blessed and thanked by all. The following morning, a smudging ceremony is also often performed. It is remarkable that at the moment of slaughter, the cow presents herself in a gesture of sacrifice as if to say she understands the need far better than any of the human participants in the event. It is to be noted that Steiner is said to have described that the only time that a member of the animal kingdom can express I of itself is at the moment of death. So that our readers may also share in our "tradition," we present the essential words gifted by Steiner in the following article "Honoring the Cow."

One of the stumbling blocks for the sensitive individual when coming to biodynamic agriculture is the requirement to use various animal parts, particularly those from the cow, an animal held sacred in human history by many peoples and even to the present day in India by those practicing Sanātana Dharma (known as Hinduism in the West). Adherents of the animal rights movements are inclined to view any animal slaughter with particular revulsion, and strict vegetarians are often of the same view. How does the biodynamic practitioner overcome their objections to the use of the various parts that must be harvested from a dead cow, most often from one just freshly slaughtered?

One of the contributing factors to these objections to the slaughter of an animal probably lies in the present-day avoidance of any discussion of the subject of death in modern society. In the first place, it is important to recognize that nothing lives forever. As simple-minded as such a statement may be, many human beings in this day and age prefer to deny this fact, especially when applied to them. The transference of this denial is easily extended to the animal kingdom when we have been conditioned to view the human being merely as a more advanced animal, thanks largely to Darwin's theory of evolution as it plays out in our current system of education. It is particularly helpful to shine the light of Rudolf Steiner's spiritual-scientific research on this concern over animal slaughter. Steiner points out that it is only the human being that is possessed of an "I,"[1] and that the sense of "I" for members of the animal kingdom resides in the spiritual world in the form of a "group soul." Individual animals do not have the capacity to say "I am" and mean it the way human beings can. Animals nonetheless do have a rich life of subjective feeling, but this does not include spiritual thought itself. The death of an individual animal, or in the instance under consideration, the individual cow, is of no more concern to the group soul of the cow species, than the cutting of the hair or trimming of the nails is for an individual human being.

---

1 The German word *Ich* is often translated as "ego" in English. But the term *ego* has far too many connotations—both psychoanalytical and Buddhist—that Steiner does *not* mean at all. A healthier translation is the word "I." The word "I" can only ever be used in reference to oneself. There is no other word like it. When you say "I," you mean only you. When I say "I," I mean only me. The term *ego* simply does not capture this. —Ed.

Of much greater help when dealing with the question of slaughtering a cow is the magnificent picture of the animal kingdom, particularly the cow, that Steiner provides us in the lecture series he gave in October and November 1923, well before the *Agriculture Course* in June 1924. My first acquaintance with this lecture series goes back to its publication as *Man as Symphony of the Creative Word*; it was described to me by one biodynamic practitioner, Clifford Kurz, as volume two of the *Agriculture Course,* even though it was presented before that course.[2]

Man as Synthesis of Eagle, Lion, and Ox/Cow

Steiner examines three representative members of the animal kingdom and their relationship to the human kingdom. Of special note is his description of the debt that the human owes to the animal kingdom, with none greater than the debt that the human being owes to the cow. While I will attempt to excerpt the most meaningful statements that Steiner makes in these lectures to give a capsule view, please bear in mind that those statements will be removed from context. I would highly recommend to anyone a thorough reading of these lectures.

The first representative of the animal kingdom Steiner addresses is the bird, specifically the eagle as an archetypical representative. To gain a "true picture, for instance, of the bird's etheric body," an awakened imagination will reveal that "the etheric bird is nothing but a head." It is "Under the influence of the sun's rays" that the bird "has received its plumage." The "sun also has non-physical powers" and it is these powers that "give the different species of birds their rich and varied colors and specific form of plumage." This sun force "is the same principle that makes the human brain the vehicle for thought." Steiner further states, "The power that creates the convolutions of

---

[2] This lecture course is currently published as *Harmony of the Creative Word: The Human Being & the Elemental, Animal, Plant, and Mineral Kingdoms* (CW 230).

the human brain and enables it to take up the inner salt force that provides the basis for the faculty of thought also gives the eagle in the air its feathers." In the bird kingdom, this sun force manifests in the physical aspect of feathers, whereas in the human being, these powers "on the astral plane...give rise to thoughts."

Next, Steiner takes a look at a second representative of the animal kingdom when he considers a mammal, choosing the lion as an archetypal example. There is no other animal except the lion and those related to it that "has such a wonderful, mysterious breathing process." It is in the lion that "a kind of balance exists between breathing and circulation." Although the heaviness of the digestive process weighs down the lion's circulation, this does not happen to the tremendous degree that is found in the case of the cow.

With the lion and its short digestive tract, the entire digestive process is completed very quickly. "In lions, more than in any other animal, the inner rhythms of breathing and heartbeat are in inner balance and harmony." The lion has a ravenous appetite "because it is part of their nature that hunger causes them much more pain than it causes other animals." Lions will find "inner satisfaction in the even rhythms of their breathing and circulation." It is at the point "when the food has passed over into the blood that regulates the heartbeat," that they are "wholly lion when they experience the deep inner satisfaction of the blood beating upward and of the breath pulsing downward." Within the human being, it is the head that most resembles the bird, and it is the chest, "where the rhythms of circulation and breathing meet," that most resembles the lion. Steiner does not subsequently address the relationship of the lion to the human being in the same way as he does with the other representatives of the animal kingdom, as will be seen later.

How the Cow Supports Human and Earth Evolution

When Steiner examines a third representative of the animal kingdom, "the ox or cow," the entire form of the cow, he concludes "is in fact what I may call a complete and utter digestive system!" In examining the cow from an astral perspective, Steiner says:

> What the bird up in the heights has by way of astrality out of its astral body, something that is working, as I have said, to shape the plumage, is something the cow has taken into flesh, muscle and bone. Something that is astral in the bird has become physical in the cow.[3]

Steiner further describes the cow's digestive processes relative to astrality as follows:

> ...it is part of the digestive process in the cow to develop a wonderful astrality. The cow becomes beautiful in the process of digestion. Seen in its astral aspect, this digestive process has something infinitely beautiful. In the light of ordinary bourgeois notions, bourgeois ideas of perfection, the business of digestion is the lowest of the low. Yet one is proved utterly wrong in this once a higher point of view is achieved and one sees the digestive process in the cow with the eye of the spirit. It is beautiful, it is magnificent, and it is something of a tremendously spiritual nature.
>
> Within the human being, the one-sided "physical embodiment of a certain astral element," which develops in the digestive system of the cow, is to be found "in the human digestive organs and their continuation in the limbs."

---

3 This and following quotations from *Man as Symphony of the Creative Word*, pp. 14–20, 22–23.

The three separate configurations found in the eagle (human head), the lion (human chest), and the cow (human digestive system and the limbs) are "united into one and in harmony and balance in the human being." Steiner asks us to "contemplate these things and realize again that humanity is actually born out of the whole of nature"—that, indeed, "humanity is a microcosm...[a] synthesis of eagle, lion, and ox, or cow." Our attention is drawn to Mahatma Gandhi, and Steiner states that the "remarkable phenomenon is that he has actually retained the veneration of the cow in his rationalized Hinduism.... [Gandhi] still retains the veneration of the cow." This veneration of the cow...can be understood only when one is aware of the inner connections, when one really knows the tremendous secrets that lie in the ruminating animal, in the cow. Then we can understand why people come to venerate in the cow a sublime astrality that has, as it were, become earthly, and only in this respect lowlier.

It is through our examination of these various representatives of the animal kingdom that Steiner hopes that we can "see how will, feeling and thought can be looked for outside in the cosmos, and correspondingly in the microcosm of the human being."

Steiner continues with a further elaboration of the cow in its relation to earthly forces, especially the force of gravity:

> The cow is the animal of digestion. It is, moreover, the animal that accomplishes digestion in such a way that there lies in its digestive processes the earthly reflection of something actually

super-earthly; its whole digestive process is permeated with an astrality that reflects the entire cosmos in a wonderful light-filled way. There is...a whole world in this astral organism of the cow, but everything is based on gravity.

Every day, the cow must metabolize an eighth of her weight. This binds the cow with its material substance to the earth; yet through its astrality, it is at the same time an image of the heights, of the cosmos.

Substance and Forces—Physical and Spiritual

In the third lecture of the cycle, Steiner presents a concept that requires us to distinguish between physical and spiritual substance as well as between physical and spiritual forces. While it takes a bit of effort to wrap one's mind around this concept, it leads to a greater depth of understanding of the role that the animal kingdom, and especially the cow, plays in allowing the human being to continue to evolve. A condensed version of this concept in Steiner's own words is as follows:

> The physical matter of the earth is part of everything that exists on earth...let us call this matter the physical substance of the earth, comprising all that provides the material basis for the various earthly forms; and then let us differentiate from it everything in the universe that is the opposite of physical substance, namely spiritual substance....
>
> It is not right to speak only of physical matter or physical substance.... When we look upon what is earthly we become aware of physical substance; when we look upon what is outside the earthly we become aware of spiritual substance. Today people know little of spiritual substance.

That is why they speak of the earth being who belongs in both the physical and the spiritual world—the human being—as though he too, only possessed physical substance. This however is not the case. Steiner continues:

We only speak correctly about the human being when we regard the so-called lower part of human nature as having as its basis what is in fact spiritual substance.... The lower man actually shows us a formation in spiritual substance, and the more we go toward the human head, the more the human being is made of physical substance.... Physical substance is to be found particularly in the human head. Spiritual substance, on the other hand, is spread out in a particularly beautiful way just where—if I may put it so—man extends his arms and legs into space.... In a form such as man possesses, however, we must differentiate not only the substance, but also the forces.... In the case of forces, things are exactly the opposite. In the limbs and metabolism, the substance is spiritual but the forces are physical, for instance the force of gravity in the legs. In the head the substance is physical but the forces active within it are spiritual.... Man can in fact only be understood when he is regarded in this way, as composed of physical and spiritual substantiality and of physical and spiritual dynamics.

Spiritualizing the Earth: Humanity's Debt to the Animal Kingdom

Steiner then goes on to place this in perspective with regard to the debt that the human being owes to the world:

Someone who...knows this secret of the human being...must acknowledge himself as a tremendous debtor to the world....

He is continuously withdrawing something from the earth. And he finds himself obliged to say that the spiritual substance that as man he bears within himself during earthly existence is actually needed by the earth. When man passes through death, he should in fact leave this spiritual substance behind him for the earth, for the earth continuously needs spiritual substance for its renewal. But this man cannot do, for he would then be unable to traverse his human path through the period after death. He must take this spiritual substance with him for the life between death and a new birth; he needs it, for he would disappear, so to speak, after death if he did not take this spiritual substance with him.

Only by taking the spiritual substance of his limbs and metabolism through the gate of death can man undergo the transformations that he must there undergo. He would be unable to be subject to future incarnations if he were to give back to the earth the spiritual substance that he owes to it. This he cannot do. He remains a debtor.

A similar problem exists with respect to the head organism of the human being, and Steiner tells us:

> Because throughout the entire course of earthly life spiritual forces are working in the physical substance of the head, this head substance becomes estranged from the earth.... [Man] must also, in order to be man, continually imbue this substance of his head with extra-terrestrial forces. And when the human being dies, this is something extremely disturbing to the earth, because it must now take back the substance of the human head, which has become so foreign to it. When the human being has gone through the gate of death and yields up his head substance to the earth, then this head substance— which is now spiritualized and bears within itself what results from the spiritual—does in fact act as a poison, as a really disturbing element in the totality of the life of the earth.... If man were to take this spiritualized earth substance with him, he would continually create something adverse to all his development between death and a new birth. It would be the most terrible thing that could happen to man if he were to take this spiritualized head substance with him. It would

work incessantly on the negation of his spiritual development between death and rebirth.

One must therefore acknowledge when one sees into the truth of these things, that here too man becomes a debtor to the earth; for something for which he is indebted to the earth but has made useless for it, this he must continuously leave behind, he cannot take it with him. What man should leave for the earth he takes from it; what man should take with him, what he has made useless for the earth, this he gives over to it with his earthly dust, thus causing the earth immense suffering in the wholeness of its life, the wholeness of its being.

NATURE SPIRITS AND THE COW

In *Man as Symphony of the Creative Word*, Rudolf Steiner beautifully describes for us the relationship of the cow to the elemental beings in the following:

> *Beneath a grazing herd of cattle, the earth is quickened to joyful vigorous life, and the elemental spirits down there rejoice because they are assured of their nourishment from the cosmos through the existence of the creatures grazing above them.*

Steiner continues with this picture of the role of the cow:

> *We would see that remarkable aura of the cow, which is in such marked contrast to earthly existence because it is entirely cosmic; we would see the lively merriment in the senses of the elemental earth spirits, who can perceive what has been lost to them because they are sentenced to live out their existence in the darkness of the earth. For these spirits what appears in the cow is Sun. The elemental spirits whose dwelling place is in the earth cannot rejoice in the physical sun, but they can rejoice in the astral bodies of the animals that chew the cud.*

Honoring the Cow

It is at this point that Steiner tells us how the kingdoms of nature provide the remedies for the human failures and shortcomings outlined above. The eagle, with its feathers that have received spiritual forces from the sun and the rest of the cosmos, because it does not reincarnate, is able when it dies to give over "spiritualized earthly matter" to spirit land and it is changed back into spiritual substance.

After death, as Steiner explains, the "spiritualized physical matter of the eagle's nature flies into the distance in order to unite itself with the spiritual substance of spirit land." In addressing the contribution of the cow, Steiner points out the cow's similarity to the human metabolic system and further enlightens us as to the significance of the cow:

> The cow is the animal of digestion. And strange as it sounds, this animal of digestion consists essentially of spiritual substance into which the physical matter consumed is merely scattered and diffused. There in the cow you have the spiritual substance... and physical substance penetrates here everywhere and is absorbed, digested by the spiritual substance. It is in order that this may happen in a really thorough way that the process of digestion is so thorough and comprehensive in the cow. It is really the most thorough digestive process that can be conceived, and in this respect—if I may put it so—the cow is really most thorough in the business of being an animal. It is thoroughly animal and actually brings animal nature—this animal egoism, this animal egoity—down to earth from the universe and into the sphere of earth's gravity.
>
> No other animal has the same relation between blood weight and total body weight as the cow; other animals have either less or more blood than the cow in proportion to the weight of the body. Weight has to do with gravity and the blood with egoity; not with the "I," for this is only possessed by man, but with egoity, with individual existence. The blood also makes the animal an animal—the higher animal at least. And we might say that the cow has solved the world problem as to the right relation between the weight of the blood and the weight of the whole body—if one wishes to be as thoroughly animal as possible.

The individuality of an animal rises only to how it *feels* subjectively. The animal cannot *think* of a universal ethic of doing unto others as they would have done unto them. Animals live under the law of an eye for an eye, and are kind to those who are kind to them. When animal interests predominate in the human being, we call this egotism. But the human I is beyond how something affects my interests subjectively and instead is loyal to how I can benefit the whole as an attitude of devotion to others. Steiner then reminds us that the cow with its "twelfth part of body weight in the weight of her blood" bears an exact relationship to the twelvefold division of the zodiac, which the ancients referred to as 'the circle of animal figures.' All other animals have a different proportion of blood to body weight, and for the human being, "the blood is one-thirteenth of the body weight." Steiner tells us: "You see, the cow aims to express the whole of animal nature in terms of weight, to bring something cosmic to expression in the most thorough way possible.... Everything to do with the cow is of such a nature that the forces of the earth are worked into spiritual substance."

Finally, Steiner identifies for us the contribution of the cow to the future of the earth:

> When the cow dies, the spiritual substance that it bears within it can be taken up by the earth together with the earthly matter to benefit the life of the whole earth. And man is right when he feels in regard to the cow: You are the true beast of sacrifice, for essentially you continually give to the earth what it

needs, without which it could not continue to exist and would harden and dry up. You continually give spiritual substance to the earth and renew the inner mobility, the inner living quality of the earth.

Steiner goes on to speak of the "remarkable contrast" between the eagle and the cow:

> ...the eagle, which, when it dies, takes away into the expanses of spirit land the earth matter that has become spiritualized and therefore useless for the earth; and the cow, which, when it dies, gives heavenly matter to the earth for its renewal. The eagle takes from the earth what the earth can no longer use, what must return to spirit-land. The cow carries into the earth what the earth continually needs as forces of renewal from spirit-land.

Placing this awareness into perspective, Steiner says,

> And now you will wonder even less that a religious world conception that penetrates so deeply into the spiritual as does Hinduism venerates the cow, for it is the animal that continually spiritualizes the earth and continually gives to the earth that spiritual substance that it has taken from the cosmos.

When the biodynamic practitioner approaches the harvesting of the various preparation sheath materials from the cow at slaughter, it needs to be done with total reverence toward the incredible significance the cow has for the entire earth, which we are given in this magnificent picture that Rudolf Steiner places before us.

In using the several parts of the cow when we make the biodynamic preparations, we are, in a real sense, entering into the role of a priest in bringing the spiritual substance of the cow to an even higher state of being in the final product of the biodynamic preparations. Alex Podolinsky has described the making of the preparations as "the only true sacrament" in the world today.

On the Harvesting of the Preparation Sheaths

When the biodynamic practitioner is confronted with the objections raised by some vegetarians or those immersed in the animal rights movement, it may be helpful to remind them that if one completely removes the animal kingdom from its contribution to agriculture, then there is a substantial danger that our food supply will disappear. Some years ago, Maria Thun was challenged by a vegan vegetarian acquaintance to grow food organically for him but without using anything derived from an animal source. That is, no animal manure could be used on the growing beds, nor could any of the biodynamic preparations be used other than stinging nettle (BD504), valerian (BD507), and Equisetum arvense (BD508). Vegetable matter compost was, of course, available, but it was not made with any of the biodynamic preparations whose production involved an animal sheath. An adjacent area growing similar crops was treated biodynamically.

Interestingly, Maria Thun approached this problem from a more complete and holistic scientific perspective by using seeds saved from each of these plots to plant back on the plots in the following years. The net result of this experiment was that by the end of the third year, some of the seeds from the "vegetarian" plot had lost viability, and by the end of the fifth year, the experiment had to be ended because most of the seed from the animal-free plot was no longer capable of producing a crop.[4]

---

[4] Description of this experiment was found in an earlier edition of Maria Thun's *Working with the Stars*, but the exact issue could not be located at this time.

Honoring the Cow

If one tries to separate the animal influence from the growing of our food, then the consequences may well be that in a short time we will not be able to grow food. Of course, the vegetarian "purist" can still grow crops organically, but only by the hypocrisy of obtaining seeds from a source that does not exclude the animal from its seed production. Such an approach would allow one to continue their purist approach but only by denying the reality of the situation.

In the light of what Steiner says of the sacrifice of the cow and the need that the earth may have for the spiritual substance it receives upon the death of the cow, one could wonder that there might not have been  a higher spiritual purpose accomplished by the tremendous slaughter of cattle that took place as a consequence of the threat of hoof and mouth disease as well as mad cow disease in England and Europe a few years ago. Does the animal kingdom undergo these sacrifices because we as human beings are not doing our job of "spiritualizing the earth" through the use of the biodynamic preparations? Are we looking at something similar in the massive poultry slaughters associated with the fear of bird flu? Is such sacrifice necessary because spirit-land is starving since we human beings can provide it only with the "junk food" of our materialistic thinking?

The Scientific Basis of Biodynamics

John Bradshaw

Applied Biodynamics, *no. 75, 2011–2012*

THE WASHING OF THE FEET

I give you thanks, cold, silent stone,
And bend me down in awe before you.
From you the plant in me has grown.
I give you thanks, green grass and flower,
And stoop in reverence before you.
You let me win the beast's swift power.
I thank you all, plant, beast and stone,
And bow in gratitude before you.
You led, all three, to me alone.
We give you thanks, bright child and star,
And kneel us down in love before you.
For—because thou art, we are.
Thanks flows from all the gods and lands,
And from each god again expands.
In thanks all being joins its hands.

CHRISTIAN MORGENSTERN
(trans. by Arvia MacKaye)

Rudolf Steiner, in giving his 1924 lecture series, Agriculture, which formed the foundation of biodynamics, was insistent that all his suggestions for a renewal of agriculture be thoroughly scientifically tested. A rigorous scientific approach has been the basis of biodynamic development since then. Some scientists criticise biodynamics on the basis that it is unscientific, or that it is based on mysticism, some claiming that there is no scientific evidence to show that it is any different from organic agriculture.

An example of this is the article "The Myth of Biodynamic Agriculture: Biodynamics is a scientifically sound approach to sustainable management of plant systems" by Professor Linda Chalker-Scott.[1]

In this article, Chalker-Scott states, concerning the biodynamic preparations: "These processes were not developed through scientific methodology, but rather through Steiner's own self-described meditation and clairvoyance.... The rejection of scientific objectivity in favor of a subjective, mystical approach means many of Steiner's biodynamic recommendations cannot be tested and validated by traditional methods. In practical terms, this means any effect attributed to biodynamic preparations is a matter of belief, not of fact." Further, "Given the thinness of the scientific literature and the lack of clear data supporting biodynamic preparations, it would be wise to discontinue the use of the term 'biodynamic' when referring to organic agriculture." And "The onus is on academia to keep pseudoscience out of otherwise legitimate scientific practices." It is a sad indictment against the standards of contemporary science that such ill-informed comments can be written by a scientist of her academic standing.

Inspiration in Science

Regarding the process of scientific enquiry, Fred Hoyle[2] and Raymond Arthur Littleton[3] wrote (1948):

> It is often held that scientific hypotheses are constructed, and are to be constructed, only after a detailed weighing of all possible evidence bearing on the matter, and that then and only then may one consider, and still only tentatively, any hypotheses. This traditional view, however, is largely incorrect, for not only is it absurdly impossible of application, but it is contradicted by the history of the development of any scientific

---

[1] Linda Chalker-Scott, PhD, Extension Horticulturist and Associate Professor, Pyallup Research and Extension Center, Washington State University, Sept. 2004.

[2] Fred Hoyle (1915–2001), mathematician and physicist, Professor of Astronomy at Cambridge 1948–1973.

[3] Raymond Arthur Littleton (1911–1995), British astronomer.

theory. What happens in practice is that by intuitive insight, or other inexplicable inspiration, the theorist decides that certain features seem to him more important than others and capable of explanation by certain hypotheses. Then basing his study on these hypotheses, the attempt is made to deduce their consequences. The successful pioneer of theoretical science is he whose intuitions yield hypotheses on which satisfactory theories can be built, and conversely for the unsuccessful (as judged from a purely scientific standpoint).[4]

It matters not the source of the inspiration, the wild leap of imagination that may lead to new theories. No rational scientist would say: "I don't know where that promising idea came from; I'd better not investigate it." It is the application of the scientific method, to prove or disprove an idea or theory, that is important. Throughout history, men and women have had ideas that seemed impossible to their contemporaries. Often these ideas came as if from outside the mind, in flashes of inspiration.

Consider Friedrich August Kekulé's[5] inspiration, which led to the discovery of the molecular structure of organic compounds:

> One fine summer evening I was returning by the last omnibus through the deserted streets of the metropolis [London], which are at other times so full of life. I fell into a reverie, and lo! the atoms were gamboling before my eyes! Whenever, hitherto, these diminutive beings had appeared to me, they had always been in motion, but up to that time, I had never been able to discover the nature of their motion. Now, however, I saw how, frequently, two smaller atoms united to form a pair; how a larger one embraced two smaller ones; how still larger ones kept hold of three or even four of the smaller, whilst the whole kept whirling in a giddy dance. I saw how the larger ones formed a chain....

---

4 "The Internal Constitution of the Stars," in Occasional Notes of the Royal Astronomical Society, 1948.

5 Friedrich August Kekulé (1829–1896), founder of the theory of chemical structure.

The Scientific Basis of Biodynamics

Reaching home, Kekulé spent the night sketching the figures the atoms danced in his dream. The patterns eventually became the formulas for organic compounds. In Ghent, Kekulé dreamed again, dozing off while thinking about the formula for benzene. This time he saw chains of atoms dancing like snakes: "One of the snakes had seized hold of its own tail, and the form whirled mockingly before my eyes." The picture of the snake swallowing its own tail gave Kekulé the idea of what chemists now call the "benzene ring."[6]

Kekulé's dictum was that inspiration is a perfectly normal part of scientific investigation, but must be followed up by careful development of a theory and experimental verification. Compare Kekulé's inspirational discovery with that of Alex Podolinsky, who was giving an early biodynamic introductory lecture to farmers,[7] when: "momentarily stunning, exactly between the audience and myself, a moving, see-through picture arose: Soil; plant with biology and liquids in action and flowing up the roots and leaves; Sunlight raying down and crystallizing into leaves and downward—i.e. a motion picture of total earth, plant, cosmos happening and, through it, the audience still visible in the background. I stopped, breathlessly looked, but then had to continue the lecture. I did not seek such a motion picture...but, I was very concentrated and an inspirational overview of this important functioning Reality was received. It took a further three years...to achieve the objectivity portrayed in Lecture 1. This was undertaken by the holistic overviewing of all factors—by thinking."[8]

Whatever the source of Rudolf Steiner's inspiration, he appears to have had considerable insight into the inner workings of nature. The detailed suggestions he made for a renewal of agriculture, including the seemingly strange suggestions for the making of the biodynamic

---

6 Prescott, *Modern Chemistry*, London, 1932.
7 Podolinsky, *Bio-Dynamic Agricuture Introductory Lectures*, vol 1.
8 Podolinsky, *Living Knowledge*, p. 33.

preparations, imply some sort of inspirational source. Whether his ideas had any basis in fact would, properly, later be tested by scientific research. Ehrenfried Pfeiffer wrote: "He never proceeded from abstract dogma, but always dealt with the concrete given facts of the situation. There was such germinal potency in his indications that a few sentences or a short paragraph often sufficed to create the foundations for a farmer's or scientist's whole lifework."[9]

Scientific Validation of Biodynamics

To accuse Rudolf Steiner of rejecting scientific objectivity is simply ignorant. Steiner had the greatest respect for scientific objectivity. As a high school student, he questioned Newton's theory of color, and, while a tertiary student at the Vienna Institute of Technology, began conducting experiments on color and light, writing several papers on his conclusions. He later discovered the scientific work of Goethe, including his theory of color, and found that Goethe had conducted similar experiments to his own. Steiner repeated many of Goethe's experiments, confirming his findings, and agreeing with Goethe's conclusions. He eagerly studied all of Goethe's scientific writings, and so impressed his professor that at the age of twenty-two (1886), he was entrusted with the job of editing Goethe's scientific works for publication in the German National Literature series. Steiner gained his PhD in 1891 from the University of Rostock, his doctoral dissertation being published as a book under the title *Wahrheit und Wissenschaft (Truth and Science)*.

When Steiner met the nineteen-year-old Ehrenfried Pfeiffer in 1918, he was a university student. Steiner encouraged him to concentrate on studying the sciences, including physics, chemistry, and botany. He valued scientific qualifications and wanted scientists to thoroughly test his recommendations for agriculture. The first test of Steiner's agricultural suggestions was made in 1923, when

---

9   Steiner, *Agriculture* (UK ed.), preface, p. 6.

Dr. Pfeiffer and Dr. Wachsmuth buried cow manure in cow horns at Arlesheim, Switzerland. The horns were dug up in early summer 1924 and had been converted into the moist colloidal, concentrated microbial substance we now call BD500 (horn manure), as predicted by Steiner's theory. After the 1924 lecture series, Steiner entrusted Pfeiffer with refining the method of making and applying the biodynamic preparations he had described, including correct application rates, storage methods, etc. Pfeiffer carried out this research work meticulously and provided his results to those involved in the practical implementation of the method.

Around the same time, Steiner entrusted a research worker in Stuttgart, Lily Kolisko, with the task of testing the validity and effectiveness of the preparations. This she did together with her husband, Dr. Eugen Kolisko, a medical doctor and lecturer in medical chemistry. Steiner had earlier asked Lily to find a method of demonstrating the activity of what he referred to as "formative forces," and she developed the method of capillary dynamolysis for this purpose. The Koliskos also carried out much detailed research on the effects of very dilute substances, and on the influence of the moon and planets on plant growth. The results of their decades of collaborative work were published in a book called *Agriculture of Tomorrow* in 1939.

They established that Steiner's suggestions for preparation making were correct. They found that, when they compared the method suggested by Steiner for each preparation with alternative methods, Steiner's suggestion was always correct. For instance, cow manure buried in a cow horn converted into sweet-smelling colloidal humus and was far more effective than cow manure buried in an earthenware pot or in a wooden box (neither of which converted into humus) next to the manure-filled cow horns.

And they found that the preparations did have a significant and positive effect on plant growth, despite the very small amounts used. A rigorous scientific development at Rudolf Steiner's request has been the basis of biodynamics since then, in those countries that followed

Dr. Pfeiffer's indications. In Australia, Alex Podolinsky thoroughly tested every new development in biodynamics (with the assistance of Andrew Sargood in the early years) at the Bio-Dynamic Research Institute. Meticulous tests were carried out over six years to evaluate the effectiveness of the stirring machines developed by Kevin Twigg for activating biodynamic preparations. BD500 (horn manure) was sprayed so accurately that each square inch of soil in the comparative plots had to receive a drop of BD500. Tests included comparative chromatography tests of the stirred BD500 liquids, plant stem, leaf, and whole plant, and evaluation of the effect of the variously stirred BD500 on soil and plants over time. Most important was the long-term monitoring of soil conversion, soil development, and biodynamic upper plant expression.

The stirring machines proved more effective at activating the preparations than expert hand stirring, for any amount of water over twelve gallons (54 liters).[10] For biodynamics to spread widely, it was essential to stir larger quantities of water at a time than twelve gallons (the soil activator spray BD500 is applied at the rate of three gallons per acre). The maximum amount of water that can be stirred in one vessel while maintaining the correct vortex and chaos characteristics was established as sixty gallons/two hundred seventy liters (twenty acres of BD500). For broad acreages, linked series of stirring vessels were developed, so far up to six sixty-gallon vessels (one hundred twenty acres per stir).

"Prepared" 500+ (developed by Alex as the only way to get the biodynamic compost preparations BD502 to BD507 out over broad acreages) was tested in a similar methodical way and proved considerably more effective than BD500 sprayed by itself. Many tests were done on soils, showing humus development, deepening color,

---

10 Expert hand stirring was slightly better than machine stirring for amounts below twelve gallons. Twelve gallons proved to be the limit to which a competent hand stirrer could go while still maintaining the required energetic, deep vortex and vigorous bubbling chaos.

The Scientific Basis of Biodynamics

and improvements in structure resulting from the application of the biodynamic preparations.

Using no inputs at all (for forty years), Alex was able to increase the organic matter in the top one hundred millimeters from 0.9% to 11.4% and from 0% to 2.4% at 1 meter in just the first six years of applying the biodynamic preparations. In 1989, Alex asked a Victorian Agriculture Department senior agronomist, Peter Medling, to investigate the amount of carbon dioxide locked up in his soil over the six-year period (organic matter holds CO_2). The investigation was done by Peter Medling, together with Agriculture Department scientists John Stewart and Graeme Savage, using Professor Leper's methodology.[11] The finding was that this soil had locked up a staggering 1,614 tons of carbon dioxide per hectare over the six years.

Various researchers have compared aspects of biodynamic farms in Australia with those of conventional farms:

- J. A. Lytton-Hitchings studied the physical and chemical properties of adjacent biodynamic and conventional dairy farms in Victoria.[12] He found that the biodynamic soil had greater macro-porosity to a depth of at least 420 millimeters, weaker soil (i.e. better structured) at 60, 120, and 200 millimeters, smaller, dry bulk density values at 120 and 200 millimeters, greater air-filled porosity at 200 millimeters, smaller volumetric water content during summer to a depth of 1.4 meters, and greater organic matter content in the upper 50 millimeters. He concluded: "These more favorable soil properties of the biodynamic soil have the potential to allow faster infiltration, less surface runoff, less waterlogging, deeper soil exploration by plant roots, and a longer interval between irrigations." (In fact, the interval between summer irrigations on the biodynamic

---

11 University of Melbourne.
12 J. A. Lytton-Hitchins, A. J. Koppi, A. B. McBratney, "The soil condition of adjacent bio-dynamic and conventionally managed dairy pastures in Victoria, Australia," in *Soil Use and Management*, 10, pp. 79–87.

farm was nearly twice as long as that on the conventional farm.)
- C. B. Parker and S. Cock also found better soil structure on biodynamic farms.[13,14]
- In 1991 the Dairy Research and Development Corporation funded a research project to assess biodynamic methods of dairy production based on biological and economic data. This study compared seven (later expanded to ten) matched pairs of biodynamic and conventional dairy farms in Victoria. The project was overseen by Doug Small (senior Victorian Agriculture Department Soil Researcher) and Dr. John McDonald (Victorian Agriculture Department Veterinarian). The results, together with three undergraduate thesis papers on phosphorus balance, physical and chemical soil properties, and soil and plant root characteristics on biodynamic and conventional farms formed the basis of a paper presented by Doug Small at the Australian Institute of Agricultural Science's Organic Agriculture Conference in 1993. Some of the results reported were: better soil structure on biodynamic farms; similar pasture composition; soil phosphorus slightly lower on biodynamic farms (which used no fertilizers); other mineral levels similar, but nitrate-nitrogen higher on conventional farms; biodynamic cows never suffered from bloat, and incidence of metabolic disorders was less; irrigation much less frequent on biodynamic farms; lower levels of selenium in conventional cows, lower levels of phosphorus in biodynamic cows; biodynamic cows were more fertile; biodynamic cows remained productive longer; grain feeding 600

---

13 C. B. Parker, The Phosphorus balance of a conventional and a biodynamic dairy farm, Undergraduate thesis, LaTrobe University, School of Agriculture, 1992.

14 S. Cock, A comparison of soil and plant root characteristics in irrigated summer pasture from two different farming systems. Undergraduate thesis, LaTrobe University, School of Agriculture, 1991.

percent higher on conventional farms; milk production was higher on conventional farms, but costs were lower on biodynamic farms; net returns somewhat higher on conventional farms, but if off-farm environmental effects were taken into account (nutrient run-off has been measured at up to thirty times higher from conventional farms, causing blue-green algae and other environmental problems) it was suggested that the total economic benefit would most likely be in favor of biodynamics.

- A study comparing biodynamic with conventional soils was carried out by Eric Frescher as part of his Bachelor's Degree in Civil Engineering at LaTrobe University.[15] The biodynamic farm studied had demonstrated a considerable reduction in salt-affected areas over five years. During the same period, the neighboring conventional farm studied had demonstrated significant expansion of salt-affected areas. The results showed that: the biodynamic soil was less compacted and more porous than the conventional soil in both a dry and wet state; water infiltration on the biodynamic soil was four times higher for the first ten minutes, and two and a half times higher after that; the biodynamic soil was capable of holding twice the amount of soil moisture, and contained twice the amount of carbon; the biodynamic soil contained numerous aggregates and was softer; the conventional soil contained very few aggregates and was harder; the biodynamic soil was five times less acid than the conventional soil; the biodynamic soil contained six times more nitrate-nitrogen in the top fifty millimeters and at two hundred millimeters contained ten parts per million nitrate-nitrogen compared with zero on the conventional farm; in dry conditions, the soils showed similar

---

15 E. Frescher, J. Russell, "A Comparison of Biodynamic and Conventionally Managed Soils: Hyden Western Australia," in *Biodynamic Growing* No. 4, (June 2005) pp. 29–33.

levels of biological activity (as measured by micro-organism respiration rates), but in wet conditions, the biodynamic soil biological activity was far higher.

Scientific Objectivity

A stumbling block for many scientists is that the biodynamic preparations are made in such a strange way, by putting manure or powdered quartz crystal in cow horns, or encasing herbs in animal organs. Scientific objectivity is replaced by disparaging emotional outbursts. Any scientist who reacts in this way, particularly without bothering to read any of the scientific studies of biodynamic methods, has left the realm of science. They are in the same position as the "scientific" contemporaries of Copernicus who ridiculed his preposterous suggestion that the Earth might move around the Sun, not vice versa. It would be a further fifty years before Johannes Kepler published his *Mysterium cosmographicum,* which supported Copernicus, and a further fifteen years before Galileo's public support for the Copernican model, supported by telescopic observations, slowly began to sway astronomers toward accepting a heliocentric model.

Consider the following procedures: take the inner mucosa of the fourth stomach (abomasum) of a young calf. Dry and clean it. Slice finely and soak for a few days in salty water to which vinegar has been added. Filter the solution. Or this: inject virus-infected material into a fertilized egg. Allow it to develop in the live chicken embryo for several days. Open the egg, harvest the virus, and purify it. Strange sounding procedures? The first is the method of isolating rennet for cheese making. The second is a method of manufacturing human vaccines. The difference between these strange-sounding procedures and the biodynamic preparation-making methods is that we now understand the processes involved in rennet and vaccine production, but don't yet understand the processes involved in making biodynamic preparations. Many theories have been proposed over time, which have taken decades or centuries to be proven. Many practical discoveries have been made, which have been of great benefit to humankind and have been used for centuries or millennia before being scientifically understood. Take the discovery of bronze, 6,000 years ago. It was discovered that if copper is melted with arsenic, a very hard metal would be formed. This was the birth of the Bronze Age. Later, it was found that tin and copper, in varying proportions, sometimes with the addition of other metals, made a better combination. Many refinements were made over time. It was not until the nineteenth century, however, that the actual processes, chemical formulae, and structures involved were finally understood.

Scientists have a long history of behaving unscientifically when their belief systems are threatened by new developments. Two recent examples: the Binning/Roher/Gimzewski scanning-tunnelling microscope, invented in 1982, produced atom-scale surface resolution. Surface scientists refused to believe this was possible and heaped scorn and abuse on the developers, who, despite this, won the Nobel prize in 1986 for their discovery. Stanford R. Ovshinsky developed amorphous semiconductors. Physicists attacked him with vitriol because it was "well known" that chips and transistors could be made only with single crystal silicon slices. He was almost bankrupted and his work was ignored until, ten years later, Japanese developers funded his work, and a great technological leap forward was made possible (writable CD-ROMs are but one result).

> Concepts that have proved useful for ordering things easily assume so great an authority over us, that we forget their terrestrial origin and accept them as unalterable facts. They then become labeled as "conceptual necessities," etc. The road of scientific progress is frequently blocked for long periods by such errors. (Albert Einstein)

Many agricultural scientists are locked into a narrow understanding of the process of plant-feeding dating back to Justus von

Liebig's discovery that plant roots can absorb nutrients only in solution (the 1840s). They accept this limited view as an unalterable fact and refuse to even consider that the Sun-warmth-directed intake of (water-soluble) nutrients from soil colloidal humus by fine feeder roots, essentially separate from the uptake of water by larger roots, as proposed by Alex Podolinsky, may present a fuller picture of the reality.[16]

Soil scientists have commented that the soil changes brought about within one year by the correct application of high-quality biodynamic preparations would take nature thousands of years to accomplish. Many photos are available that depict these astounding changes, and many professional farms and farmers can demonstrate the effectiveness of the biodynamic preparations in practice. Farmers following the plant-feeding theories of conventional soil scientists develop soils that are less well structured, less biologically active, lower in organic matter and humus, less capable of absorbing water, and less capable of holding water. They use large amounts of increasingly expensive (and, in the case of phosphate, finite) artificial fertilizers to support plant growth. Many nutrients from these fertilizers run off their farms into waterways, causing environmental degradation (let alone the sheer wastage). Their plants require protection from insects and diseases with toxic chemicals, which put the health of farmers, consumers, and the environment at risk. Their produce contains pesticide residues, contains more water and fewer nutrients, and has a shorter shelf life and less flavor. Their animals are less healthy and less fertile.

Farmers (worldwide) following the "Australian Professional Bio-Dynamic method," encompassing the plant-feeding theory espoused by Alex Podolinsky, together with the all-important biodynamic preparations, develop soils that are better structured, deeper, darker, more biologically active, higher in organic matter and humus, capable

---

16 Podolinsky, *Bio-Dynamic Agriculture: Introductory Lectures,* vol. 1, lect. 1.

of absorbing more water, and of holding more water. They use relatively small amounts of (non-water soluble) fertilizers and have very low nutrient runoff from their farms. Their plants don't require the application of toxic chemicals to protect them from insects and diseases. Their produce is almost completely free of pesticide contamination (environmental contamination deriving from conventional farming is impossible to completely avoid). Their produce has less water, more nutrients, a longer shelf life, and a better flavor. Their animals are healthier and more fertile. The inescapable conclusion is that the theory of plant-feeding underlying conventional farming is deeply flawed, whereas the theory of plant-feeding advanced by Alex Podolinsky, which underlies biodynamic farming, is supported by the practical results.

Most of the above characteristics of the two farming systems have already been supported by published research, while the rest are supported by general observations and published objective blind taste tests, but await funding for properly constructed studies. Sadly, though, pure science has been largely subverted, over the last thirty years, research funding being more and more linked to the commercial sector. Research that would enable farmers to improve soils and crops without importing large amounts of fertilizers and without using chemical sprays, is unlikely to attract funding from the agro-chemical industry. Scientists who question the status quo are often harassed, lose tenure, or lose funding. The source of scientific inspiration is irrelevant: whether it be through a dream, a flash of intuition, claimed clairvoyant ability, or even if the "fairies at the bottom of the garden" told you, the important point is the development of a theory based on the inspiration and the thorough scientific testing of it. Scientists must remain objective and open-minded. They must be mindful that scientists throughout history have quite wrongly and unobjectively ridiculed innovations that appeared strange or didn't fit with currently accepted theories but were later proven correct. Alex Podolinsky adds:

Biodynamics, by result, is capable of providing sustainable food production and sustainable nature enhancement, being a method obeying the laws of the organic design of Nature and Creation, providing food of high quality and able to supply humanity in equal measure to artificial means including the danger posed by GMOs, which have not been objectively examined as to their long-term effect. Biodynamics provides a sane future for the Earth. GMOs and nanotechnology are two totally unexamined prospects of great danger to the future of the Earth.

John Bradshaw edits Biodynamic Growing Magazine. *He studied primary school education at Toorak Teachers College and is a former farmer and biodynamic certifier for Demeter Bio-Dynamic Research Institute. He lives in Cranbourne South, Victoria, Australia.*

Reprinted with permission. Originally published in Biodynamic Growing Magazine, *Dec. 2009: www.bdgrowing.com.*

Sensitive Crystallization:
Revealing the Life Forces behind Biodynamics

An Interview: Hunter Francis and Philippe Coderey
Applied Biodynamics, no. 75, 2011–2012.

Developed by Ehrenfried Pfeiffer in the 1930s at the suggestion of Rudolf Steiner, the process of sensitive crystallization is a unique and powerful method used to illustrate and assess the life forces or energy fields behind matter. The method is simply employed with a solution of copper chloride mixed with the substance to be studied and then evaporated, leaving an array of crystals characteristic of the substance. For the past six years, Philippe Coderey, a vintner and winemaker from southern France, has been applying Pfeiffer's method to support his work in agriculture. By making thousands of crystallizations of wines, soils, biodynamic preparations, and more, Mr. Coderey has used the crystal patterns formed to interpret the energetics, organization, and complexity of a wide array of substances with stunning

Philippe Coderey with horns

results. These patterns serve as a sort of energetic "fingerprint" and can be very helpful in diagnosing and remedying problems related to vitality and formative forces. In September 2011, I had the pleasure of speaking with Mr. Coderey at Tablas Creek Winery in Paso Robles, California. He graciously shared his methods, insights, and many of his discoveries with me. Throughout our discussion, Mr. Coderey used many of his crystallizations to illustrate the principles and reactions we were discussing. We have annotated four examples in the figures shown below.

Hunter Francis: I'd like to start with a little history. Where are you from and what is your background in agriculture?

Philippe Coderey: I am originally from the Toulon region of Provence in southern France. My family was involved in viticulture and enology, so it was part of my upbringing. We grew wine grapes and made wine, and this led to my interest in winemaking. We also grew olives and did some farming.

HF: Did you study it formally?

PC: Yes, I studied viticulture and enology for five years first in Provence and then at the university in Dijon.

HF: What did you do after school?

PC: I got my first job in Switzerland, where I worked from 1983 to 1987. I worked for a small agricultural firm that sold inputs for vineyards and wineries—everything from synthetic fertilizers and pesticides to equipment. I started to have problems with my liver, however, due to the absorption of the chemicals through my skin. So, I knew I had to stop.

HF: And how were you led to biodynamics?

PC: My girlfriend at the time was involved with a Waldorf school, which had a biodynamic garden. I spent a lot of time in that garden and learned about the philosophy behind biodynamics there. It grew on me, and eventually, I realized it was a direction I needed to take. At the same time, I had always wanted to live in the United States, so when I learned that the Camphill community in Kimberton, Pennsylvania, was offering a three-year training in biodynamics, I decided to do it. I moved there in 1987 and stayed for almost ten years. The program was discontinued in 1993, but I stayed until 1996.

HF: Who taught the curriculum?

PC: Many people were involved. The farmers working in the community taught a lot of it, but we also had many visiting professors, including Hugh Courtney, Hugh Williams, Dennis Klocek, Alan York, Alex Podolinsky, and even Herbert Koepf.

HF: Some leading teachers! What did you do after that?

PC: I had wanted to go to California when I came to the U.S., but I ended up in Boulder after Kimberton. The Naropa Institute was setting up a biodynamic farm there and they needed help. Jim Barausky, also from the Kimberton program, started the farm. I worked there for eight months until my visa ran out in 1997.

HF: Did you stay in France at all before returning?

PC: Yes, I lived in Corsica for a few years, working in a biodynamic citrus orchard. Then, I went back to the mainland of France and worked for M. Chapoutier, which owned eight vineyards in southern France. I worked as a vineyard manager and winemaker for them for six years.

HF: Were the vineyards biodynamic?

PC: Yes, since the early 1990s.

HF: Would you say the French are more receptive to biodynamics?

PC: Not necessarily. They have an affinity for the traditional aspects of biodynamics. When it comes to age-old farming techniques that their ancestors used, it's something they can identify with. My great-grandmother, for instance, used to pay attention to the moon when she worked on the farm and garden. She even practiced healing by the laying on of hands, prayer, and the use of medicinal plants—as did many people of her time. When it comes to the esoteric side of biodynamics, however, the French can have as much trouble as anyone else. For myself, I tend to focus on the results of biodynamics—the health of the soil, resistance to disease, the taste of the fruit, etc. The proof is there; it works. When you think about it, there are many scientific phenomena that we can't see, yet still produce results we use every day. I'm thinking here of things like microwaves, radiation, and even cell phones. We readily accept the existence of those energies, even though we can't see them.

HF: So, how did you finally end up in California?

PC: I was contacted by a recruitment agency in 2004 on behalf of Randall Grahm of Bonny Doon Vineyards in Santa Cruz, California. He was looking for a French person with experience in viticulture, winemaking, and biodynamics. I seem to have fit the bill! Long story short, I was hired in 2005 and became their director of biodynamic viticulture.

HF: How many acres do they have under their management?

PC: It's constantly changing. In 2005, we were working with about sixty vineyards, but in 2006, Randall Grahm sold two of his brands. When that happened, we no longer worked with conventional growers and could focus exclusively on biodynamic growers. When I left last year, I think we were working with under a dozen vineyards.

HF: How was your experience there?

PC: It was great. It was a unique opportunity because instead of working with just one vineyard, I got to experience many different vineyards around the state.

HF: How does your work with sensitive crystallization fit into this whole trajectory?

PC: Well, ever since I started to study biodynamics, I wanted to study sensitive crystallization, because that's a part of biodynamics. The method was developed by Ehrenfried Pfeiffer, who

Figure 1: A highly mineral wine (top) vs. a fruity wine: The "mineral" wine was made from fruit from an old abandoned vineyard in southern France. The vineyard had essentially gone back to the wild. It'd had no pruning or fertilizer since World War I. The vines had extremely deep roots. So, the fruit from those vines expressed qualities of a landscape undisturbed by human intervention. In the crystallization, you can see a high degree of minerality, which is evident around the edges of Petri dish. This was reflected in the mineral taste of the wine—a very powerful expression of terroir. The fruity wine, on the other hand, has long, straight needles and much weaker forces around the periphery. The lack of cross-hatching of the needles indicates the wine has very little complexity. In fact, the wine can be considered a "fruit bomb." This means the wine has a very simple expression of fruitiness, but lacks the complexity and expression of terroir.

was a student of Rudolf Steiner. Steiner gave Pfeiffer indications as to how he could use crystallization as a means of demonstrating the superiority of biodynamics versus conventional agriculture. It took Pfeiffer about twenty years to develop the method. He tried using many salts in the process but eventually settled on the use of copper chloride because with copper chloride, you have the greatest diversity of expression within the crystallization. With copper chloride, you get different layers with a lot of character.

HF: So that's the method you use?

PC: Absolutely, it's crucial. Consistency is especially important in comparing different crystallizations. If you use a different technique, you will have a different result; so, it is important to use the same method, especially if you are comparing the work of other people. Also, to ensure the relevance of what I am looking at, I use controls and robust sampling of any given substance.

HF: Where did you learn the technique?

PC: I first heard about the process at Kimberton, but we didn't have a lab there. We talked about it and looked at pictures of crystallizations. There were still people around at that time who had worked with Pfeiffer and who knew about his work. It was very interesting. But it wasn't until I was at Bonny Doon that I began to practice it. When I told Randall Grahm about the process, he asked me to find someone who could teach me how to do it. Well, I did some searching and finally contacted the Pfeiffer Institute in Lyon, France. They had done some crystallizations for Chapoutier, so I asked them if they could teach me. They agreed, but I would have had to move back to France to study with them. Since I was working, I couldn't do that; so, I asked if they knew of anyone else I could study with. Ironically, they referred me to a man in Provence who turned out to be an old high school classmate of mine!

HF: Who was that?

PC: Christian Marcel. He came to Bonny Doon in 2005 and helped me set up a lab there. He started to teach me how to do the tests and how to read them. Over the following year, I would send him my crystallizations, and we would discuss them together. He came back the next year to give me more training, and it has developed from there.

HF: So what exactly is the process? Can you tell me how you do a crystallization?

Sensitive Crystallization: Revealing the Life Forces behind Biodynamics

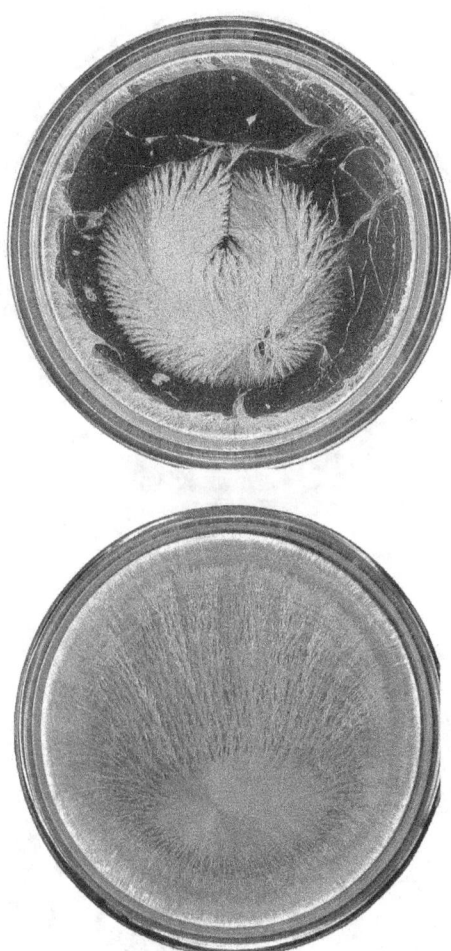

Figure 2: Poor quality BD500 (top) vs. high-quality BD500: In the first picture, of the "poor" quality preparation, it appears the preparation has weak life forces. This was the result of a preparation made in very dry soil, which had poor water retention in an area that received only ten inches of rain in the winter. This made for a very mineralized preparation. The "stronger" sample was from an area that received about fifty inches of rain in the winter. The topsoil was much deeper and had much greater water retention capacity. The result is a more powerful expression of forces, as evidenced by the way it filled out the Petri dish.

PC: Of course. First, you need a dehydration chamber, which is essentially a wooden box with two glass shelves. Mine is about two feet by two feet by one and a half feet in size. It opens in the front and has a window so you can see what's happening inside. There are three sets of shock absorbers—one under the box, and a set under each shelf—as the process must occur undisturbed by any sort of movement. There are holes in the top of the box, so moisture can exit. At the bottom, there are some small lights

Figure 3: Compost before BD preparations (top) vs. compost after application of BD preparations: Clearly, there is much better organization of the forces after the application of BD preparations. In the crystallization of the compost that had not received the preparations, you can make out several centers, which reflect the five different ingredients the compost had been made with. In the second picture (after the preparations) you can see that the centers seem to merge into one another. This indicates the ingredients are working together and is likely what Steiner meant by saying the preparations help *"vivify and organize a compost"* in his Agriculture lectures.

to warm up the box. I keep the temperature at 90°F (30° C). The Petri dishes used in the crystallization are made of Pyrex and must be perfectly flat. I have mine custom-made. The solution used in the Petri dish is a total of five milliliters: one milliliter of wine (or other solution), two milliliters of purified water, and two milliliters of ten percent copper chloride solution. I mix the solution in a beaker, and I always do a series of crystallizations.

Replication of the result is important to see. Because the Petri dishes are very hard to clean afterward, I usually do about three to four. When you are ready, you put the solution in the Petri dishes, and let them incubate for about eight to ten hours. I do mine overnight. In the morning, the crystallizations are complete.

HF: What types of crystallizations can you do besides wine? Can you do orange juice, for instance?

PC: Yes. I can do everything except oils.

HF: Can you test solids, such as soil?

PC: Yes. In that case, I take ten grams of the soil and let it soak in forty grams (milliliters) of HPLC (high-performance liquid chromatography) grade filtered water (a 1:4 ratio) for an hour. I stir it to make sure it is homogenous. Then, I filter the solution and use one milliliter of that in the crystallization recipe. (A coffee filter of appropriate porosity is used, making sure the tightness of the weave is neither too loose nor too restrictive. This can be determined through trial and error. For the sake of consistency, one should always use the same kind of filter.)

HF: What about the ten percent copper chloride solution? How do you make that?

PC: I buy the purest copper chloride possible—dry, ninety-eight percent. To make the solution, I take twenty-five grams of the dry copper chloride and dissolve it into two hundred fifty milliliters of purified water. This is according to the original Pfeiffer recipe. Everyone doing crystallizations around the world uses that recipe. You can also find it in Christian Marcel's book *Sensitive Crystallization*, which is widely available.

HF: Have you done many crystallizations with food?

PC: Not many. A few months ago, I started to crystallize tomatoes—biodynamic vs. conventional. I'd like to do more with food.

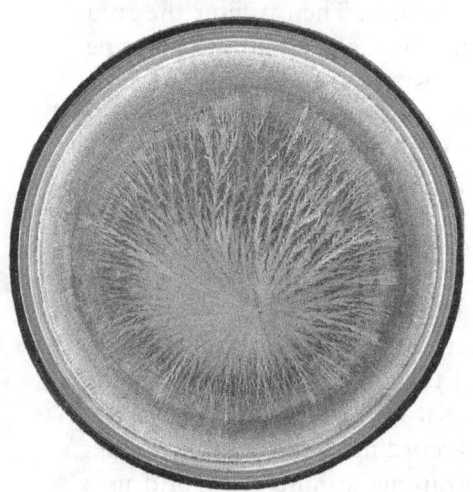

Figure 4: Soil before BD preparations (top) vs. soil after application of BD preparations: This experiment was to see if the compost preparations would have a similar effect on soils as on compost. The soil here was from a biodynamically-managed garden in California. The garden was approximately 35 feet by 35 feet. As an experiment, the compost preparations were buried in the soil as one would place them in compost. This was done in a circular arrangement of the five solid preparations, with each one placed along the periphery of a circle with a diameter of approximately 25 feet. The preparations were buried equidistant from each other. The sixth preparation (valerian) was stirred as usual and sprayed on top of the whole area. The soil treated in this manner was then compared to a sample of the same soil taken before the experiment. When one looks at the difference between the crystallizations, one can see how the preparations vivified and re-enlivened the soil—and only two hours after application! Note: The first picture (before the addition of the preparations) was taken about three days after the first rain following the Fukushima disaster. One would expect to see a much better organization of forces in BD soil, indicating the rain affected the biodynamic soil negatively. Rainwater normally vivifies soil, whereas in this case, it seems to have weakened it. After the new application of the preparations, however, we can see that the organization of the forces was enhanced.

Sensitive Crystallization: Revealing the Life Forces behind Biodynamics

HF: *What can you say about the art of interpretation, after the crystallizations are complete?*

PC: Well, it depends on the substance you are studying. In wines, for instance, you have three fundamental areas to consider: the center, where the crystallization starts, the periphery around the edges of the Petri dish, and the area in between the center and periphery. The location of the center of a wine crystallization is like the fingerprint of the varietal. A Grenache, for example, is always off-center, toward the bottom of the crystallization. A wine made with four types of grapes will exhibit four distinct centers. The center also reveals the fruit character of the wine—its fruitiness. The smaller the centers are, the less fruitiness you have in the wine. The periphery tells you about the mineral content of the wine, which relates to terroir since it comes from the soil. The relationship between the center and the periphery tells you something about the overall character of the wine, for example, whether the minerality (around the edges) is overpowering the wine. The middle regions tell you about the vegetal and floral character of the wine. The shape and arrangement of the radiating crystals tell you something about the wine. The primary or "pine needle" radiation from the center out to the periphery refers to the life forces being expressed in the wine. The secondary branching cross-hatching of the radiating needles refers to the organizational forces or complexity. So, if you have long straight lines coming out from a strong center with little cross-hatching, you have a fruity wine with weak vegetal and floral character, and not much complexity. Of course, to make these connections myself, I continually had to taste the wines I was crystallizing. This has been essential in being able to interpret the crystallizations.

HF: *What about crystallizations of the biodynamic preparations? What have you seen in them?*

PC: Overall, you can see in the crystallizations of the preparations (and soils or compost that have had preparations applied) that the preparations do a lot to help organize the crystals in the Petri dishes. Crystallizations showing the effect of the preparations also tend to have a more distinct center. You know, when I first read Steiner's *Agriculture,* I didn't know what he meant when he said the preparations "organize the compost." Now I can see what he means in the crystallizations. You can also see changes over time. Take the case of BD500 (horn manure), for instance. The process involved in that preparation has a lot to do with digestion. We as humans say, "I am." A cow would say, "I digest." That is the essence. When you place a manure-filled horn in the soil, you should realize the soil also digests, thus enhancing the process begun in the rumen of the cow. The digestion continues through the wintertime, in the soil, as it is digesting and transforming its organic matter. The next year, you spray the digested material on the soil, thus reinforcing the process. One thing that I've seen is that when you look at a soil that has had the preparations applied over several years, the quality of the crystallizations progressively improves.

HF: Have you experimented at all with other preparations?

PC: I've mostly done crystallizations with BD500 and barrel compost (sold through JPI as Biodynamic Compound Preparation), but I've experimented with BD501 (horn silica). I've also crystallized soils and composts before receiving the preparations and after receiving the preparations to see what changes occurred in the organization of the crystallizations.

HF: Is there anything else you should consider in looking at crystallizations done for soil or compost?

PC: One thing that can have a marked effect is the quality of the water used in agriculture. People sometimes don't think of this in making preparations or compost. Irrigation water, if at all contaminated, can seriously interfere with the quality of the crystallization, as opposed to purified water. One thing I've noted, however, is how big an influence passing irrigation water through a flowform can have on changing its energy field. Flowforms help organize the forces reflected in crystallizations.

HF: Are there any other applications you'd like to mention?

PC: The potential range of application is really astonishing, and the sensitivity of the crystallizations to energetic effects seems to be far-reaching. I've been inspired by the work of Masaru Emoto in Japan, who looks at crystallizations of water that has been frozen. At one point, I decided to emulate his work, but on dry soil rather than on water. In the experiment, I dried the soil and stored it for a long time. Before testing it, I divided the soil into five equal portions—the same soil, mind you—and placed each sample over pieces of paper with five different words on them ranging from vulgar insults to "gratitude." Each one yielded a different crystallization, even though they had come from the same soil. And while they had similarities indicating that they were from a similar material, they each yielded a different qualitative result, with gratitude being the most beautiful in terms of the organization of the crystals. What this tells me is that soils, like water, are sensitive to many stimuli—including our intentions. It's interesting to note that Steiner first introduced the idea of doing crystallizations during the time he was offering consultations to physicians. In that role, he had the opportunity to visit many patients. As he was working, he noticed frost crystals that had formed on the windows of the rooms of the patients. What captured his interest, however, was that similarly shaped crystals appeared on the windows of patients who were afflicted with the same diseases! There is, no doubt, an artistic, qualitative aspect to interpreting sensitive crystallizations. The more you do them, however, the sharper your intuition becomes. It is fascinating work.

HF: *Indeed it is! Thank you very much for sharing your knowledge, Philippe.*

Philippe Coderey comes from a long line of vine growers and farmers in southern France and Switzerland. His family name finds its roots in the old French verb codurer—*meaning to cultivate vineyards—and his ancestors were named for their vocation during the 11th century. He grew up on the family vineyard in Provence, France, and was fortunate to be exposed to traditional viticulture techniques through elders of the family.*

500: The Foundational Horn Manure Preparation

Hugh J. Courtney
Applied Biodynamics, *no. 5, 1993*

If one were to take a "cookbook" approach to the making of BD500 the following recipe might be all one needs to say on the subject. But as anyone who has ever experienced making BD500 knows, such a recipe is deceptively simple and the end product may not necessarily turn out to be "decent" BD500. To include all the possible nuances and variations involved in making "good" BD500 would perhaps require volumes, so I shall attempt here to cover some of what my own experience suggests may be the most important considerations to take into account.

Ingredients:

- 1 to 2,000 cow horns as appropriate for needs of the "agricultural individuality."
- Enough cow manure to fill the above number of cow horns. Variable.
- Bury in good soil during the winter months.
- Dig up in spring when manure in the horn is no longer "green."
- Each horn can be expected to yield approximately 2½ units of finished BD500 on average.

The first and foremost lesson I learned from Josephine Porter, who served as my mentor in making the biodynamic preparations, was to

500: The Foundational Horn Manure Preparation

"return to the source." So, the first step in making BD500 each year is to reread what Rudolf Steiner had to say:

> We take manure, such as we have available. We stuff it into the horn of a cow, and bury the horn a certain depth into the earth say about 18 in. to 2 ft. 6 in., provided the soil below is not too clayey or too sandy. (We can choose a good soil for the purpose. It should not be too sandy.) You see, by burying the horn with its filling of manure, we preserve in the horn the forces it was accustomed to exert within the cow itself, namely the property of raying back whatever is life-giving and astral. Through the fact that it is outwardly surrounded by the earth, all the radiations that tend to etherealize and astralize are poured into the inner hollow of the horn. And the manure inside the horn is inwardly quickened with these forces, which thus gather up and attract from the surrounding earth all that is ethereal and life-giving.
>
> And so, throughout the winter in the season when the Earth is most alive the entire content of the horn becomes inwardly alive. For the Earth is most inwardly alive in wintertime. All that is living is stored up in this manure. Thus, in the content of the horn, we get a highly concentrated, life-giving manuring force. Thereafter we can dig out the horn. We take out the manure it contains.[1]

Aside from responses to certain questions pertaining to specific facets of the making of BD500, which are very interesting in themselves, Rudolf Steiner limits his description to the above two paragraphs. Once again, that description is deceptively simple.

We prefer to gather fresh cow manure from the pasture at a time of the year when the cows are receiving at least half of their forage as hay. We prefer to use only cow manure and attempt to avoid any bull, heifer, steer, or calf manure even to the point of segregating the cows concerned in their own separate pasture. Ideally, the manure should also be from lactating cows. We subsequently screen the manure through ½ inch mesh hardware cloth screening packed

---

1 Steiner, *Agriculture*, p. 74.

tightly into buckets or wash tubs and then fill the horns. This is, of course, rather different than taking "manure, such as we have available." When filling the horns, we use large spoons, usually stainless-steel serving spoons or wooden cooking spoons, and attempt to press the manure as tightly as is possible into the horn. For this, of course, drier manure is more appropriate and this is the reason for having hay as a large portion of the cow's diet. Our justification for packing so tightly is based on Steiner's answer to one of the interesting questions mentioned above:

> *Question: Where can one get the cow horns?*
>
> Answer: It makes no difference where you get them from, but not from the refuse yard. They must be as fresh as possible. However strange it might sound, it is a fact that life in the Western hemisphere is quite a different thing from living in the Eastern hemisphere. Life in Africa, Asia, or Europe has quite another significance compared to life in America. Possibly, therefore, horns from American cattle would have to be made effective in a rather different way. Thus, it might prove necessary to tighten the manure rather more in these horns to make it denser, and hammer it more tightly.[2]

Another technique for packing tightly is to tap the tip of the cow horn firmly against a hard surface (say a concrete floor) several times during the filling process. Use the technique that is most comfortable and effective for you.

We have also come to believe that the spot where the horns are buried needs to be carefully chosen, with particular emphasis on "good" soil. The site for the horn pit definitely needs to receive treatment with the biodynamic generous treatment preparations and if the soil is of poorer quality, some well-aged biodynamic compost needs to be incorporated. We have also begun to protect the cow horns from any possible detrimental effects due to radioactivity by

---

2   Ibid., p. 79.

using a two-to-six-inch layer of milled peat moss as an immediate covering followed by the fertile soil dug from the pit.[3] Repeated use of the same site seems to produce an improvement of the final BD500 product in each successive year.

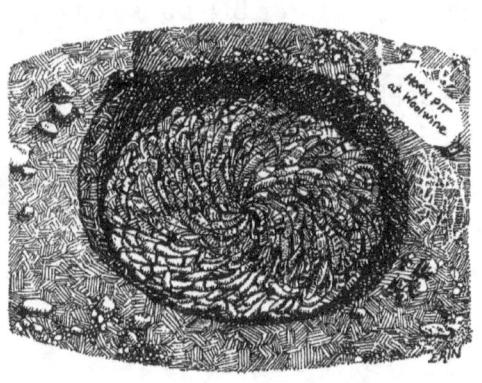

The digging up of the horns is a much-debated question, with times ranging from the very first moment after the spring equinox all the way to digging it up when you are ready to use it, even if that is the Fall of the year (as is covered in another question-and-answer exchange with Steiner and the audience at the agriculture course). Our own preference on timing is to bury the BD500 horns before Advent, if possible, but certainly before Winter Solstice, and to dig them up on Ascension Day or shortly thereafter. Every year has its own requirements.

The reader should also study what well-qualified preparation makers have had to say on the subject. Among those people, one can commend to your attention Evelyn Speiden Gregg, Lily Kolisko, Maria Thun, and Alex Podolinsky.

Even though the Josephine Porter Institute for Applied Biodynamics, Inc., derives a fair portion of its financial support from the sale of units of BD500, we can only encourage everyone to begin learning to make their own preparations. (If you have difficulty obtaining cow horns, the Josephine Porter Institute has a plentiful supply for sale. Call or write for prices.) The "recipe" for making BD500 is a gift from spiritual worlds through the agency of Rudolf Steiner and the knowledge of how to make "proper" BD500 needs to be widely

---

3   Use the peat moss suggested in studying the work of Ehrenfried Pfeiffer.

shared as a spiritual necessity in the world today. Besides the spiritual dimension involved, one's understanding of the various preparations and how they should be used somehow seems to increase substantially the more one makes the preparations. Understanding and awareness of BD500 is acquired not through the intellect but through the will in the act of stuffing manure into the cow horn. Try it, you'll like it!

Further Thoughts on Making BD500

Hugh J. Courtney
Applied Biodynamics, *no. 9, 1994*

In the Michaelmas issue of *Applied Biodynamics* last year, in the article entitled "Recipe for Making BD500," I outlined the bare essentials of placing the manure in the cow horns and burying them in the earth. As anyone knows who has ever gone through the process, there is much more to the making of BD500 than this. Crucial to the successful making of BD500 is its excavation and its subsequent evaluation. Perhaps even more crucial may be the attitude with which one approaches the entire task of making and digging up the highly precious horn manure.

Let us look first at the task of excavation and evaluation of this preparation. The most obvious, but often overlooked, requirement is to have marked the horn pit carefully and thoroughly enough so that you can find it again when you are ready to dig up the pit. I can well remember one of my first experiences at making horn manure at a friend's place in Eastern Virginia. I took careful note of just how far it was from the fifth fence post from the corner, in full confidence that it would be an easy matter to recover the three or four horns that I was burying. Imagine my dismay when I made my return visit the following spring to discover that over the winter months, my friend's farmer neighbor, who was responsible for maintaining the fence, had removed it, bulldozed the entire fence line, and was in the process of reconstructing it. Needless to say, the whereabouts of that particular horn pit remain hidden to this day.

Assuming that you have full assurance of the exact location of your buried horns, the excavation should proceed in a manner so as to remove all soil from atop the horns, while trying to disturb the horns themselves as little as possible. That is, when there is still considerable soil covering some of the horns, it is quite easy to be so rough in the digging up that you can actually spill dirt into the opening of the horns that happen to be facing in the direction of the soil being moved. I begin the actual step of lifting and removing a cow horn from the pit by scraping off the excess soil that might be clinging to the exterior of the horn. For this, I use a small trowel, a kitchen or putty knife, or even a pocket knife. Next, any soil that may have lodged in the "throat" of the cow horn is carefully removed with the blade of a pocket knife, until I am reasonably certain that only horn manure is still contained in the horn. When all the horns are taken from the pit in a similar fashion, I then begin the process of actually emptying the horns themselves. Often, the first step is to smell the manure in the horn, hoping that my nostrils will detect an earthy smell, rather than any odor resembling the raw cow manure that was packed into the horns the previous Fall. Almost simultaneously, my eyes are carefully assessing the color of the material in the horn. Most desired is to see a deep rich black color, followed next in preference by an earthy brown color. Anything that resembles the greenish color of the raw manure that went into the horns to begin with, particularly when coupled with a raw manure smell, is cause to set that particular horn aside. However, if the color is wrong, but the smell is right, the problem most likely is a lack of air, and the manure from such a horn if emptied and exposed to air for as little as 24 hours will usually take on a more acceptable color. (Often, a major cause of this lack of air is the water that has somehow entered the horn, in spite of all efforts to position the horn in the pit to prevent this.) If it happens that you dig up your horns quite early in the Spring, and you experience either a wrong color or odor or both, my best recommendation is to rebury the horns. Some people who contacted me this Spring with this problem and followed my recommendation reported better results when reexamining the pit as little as two weeks later. In some instances, with horns that do not pass muster, either by smell, sight, or both, storing them in our root cellar for a few weeks has led to a "curing" process that yields BD500 with good smell and color. This root cellar location is well below the normal soil level, and serves, in essence, as a kind of horn pit in and of itself. The lack of soil actually surrounding the horns constitutes the major difference between leaving the horns buried, or allowing them to "cure" in the root cellar. During this past winter, many areas of the country experienced long periods of heavy ice. This may have resulted in much less "breathing in" of air into the Earth. A snow cover does not seem to limit the air penetration in the same way that the ice cover appears to do. However, a number of people who reported problems with their BD500 were in a location that did not experience any snow or ice this past winter. My speculative explanation for this is that we were experiencing a "light/air/flower" year. That is, the major area for the working of the formative forces for the year was in the realm of the air, with little or no penetration below the soil surface during the winter. Since that speculation continued the fire-earth-air-water rhythm that I have decided I have been observing over the last several years, I have a certain fondness for it. Such speculation requires far more than just a few years of observation before it can be deemed to have substance. Also, I have not been too confident in determining the beginning of each year, or at least, the change-over-time of one formative force or elemental impulse into its successor. Currently, my choice for such a change-over point would be St. John's Day. If you feel a certain resonance of thinking as regards my speculation, I would be very glad to hear your comments.

There is, however, perhaps an even more critical factor in assuring the production of acceptable BD500. That factor involves the attitude or attentiveness that one brings to the horn stuffing process. In my effort to identify and label this factor, I have referred to it in my own mind as the "green thumb/black thumb syndrome." More than once over the years, I have seen different people achieve different results with the finished product of BD500, in spite of using manure from the same source, horns of equivalent value, and the horns buried side-by-side in the same pit. Initially, I preferred to assign the difference to some innate, but not easily observed difference in the quality of the horns. Recently, however, the evidence of a "green thumb/black thumb" factor has become too strong to ignore any longer. Accordingly, in the future, each individual participating in our horn stuffing activities will either bury his/her horns in a separate pit or else use a separate individually assigned marking on horns buried in a common pit. In the Spring of the year, when the pit is dug up, the resultant BD500 will be indicative of success or failure. Any "grade" is actually assigned by the "elemental kingdom," and I am not by myself or anyone else overseeing the operation.

Let us examine more closely this factor of attitude or attentiveness in the making of not only BD500 but all the other preparations as well. During his visit to Kimberton, Pennsylvania, in June 1992, Alex Podolinsky was heard to say that the biodynamic preparations are the only true sacrament on the Earth today. Whether you agree with his statement or not, I would believe that it is at least indicative of the attitude one must bring to the preparation-making task. It was certainly the attitude held by Josephine Porter who was my mentor in the preparation-making art. Whatever success I have subsequently enjoyed in that art, I must attribute to my emulation of her attitude regarding the sacredness or holiness of this task of preparation-making. If we approach the making of the preparations as just one more farm task that we have to get done, and without any consciousness of their importance in healing the Earth, we should not be too surprised if we achieve rather indifferent results. One of our directors here at JPI, Jim Marquardt, has suggested that an appropriate ritual should accompany the making of each preparation. While we shall certainly consider the feasibility of this, I hesitate to do so in order to avoid "fixing" or "dogmatizing" the process. What is really needed is an "inner ritual" by each person participating in the preparation-making activity. Perhaps in the future, one could consider a "sweat lodge" or some other "purification" process, but the whole biodynamic effort requires the individual to be constantly working on inner purification. Whatever might be stated here as regards the making of the preparations should be seen as equally applicable to the use of the preparations. If one is in an argumentative or otherwise "bad" frame of mind, one should probably plan to make the preparations another day. If you find yourself performing the task of just stuffing another cow horn routinely without really concentrating on the need to pack it as tightly as possible, then it would be advisable to stop and refocus your concentration. While I trust no one will take my next comment amiss and assume it to be a chauvinistic statement, it is probable that any woman experiencing menstruation should avoid making any preparations during that time. Such abstinence is observed in anthroposophic pharmacies in Europe when potentizing remedies. The making of a medicament for the healing of the Earth, such as BD500, could be affected by the very strong connection to the forces of the Moon usually associated with menstruation. Just to put a different perspective on this question, I should say that it has been my observation that women generally tend to produce better biodynamic preparations than men. A man probably needs to work very hard every day of the month to bring a balance into this preparation-making process, while a woman brings a natural affinity or connection to the process except for those few days of the month when the Moon energies may interfere. In any case, this question at least needs to be placed on the table as a possible factor.

Of more direct concern is ridding ourselves of an attitude that regards the preparations as just a kind of organic fertilizer that is much better than any chemicals, and somehow a bit better than most organic soil amendments. At the end of this article, the reader will see a reference to several books, which may seem quite unconnected to either agriculture or the biodynamic preparations. However, if you read any of those titles while holding in mind an awareness of the biodynamic preparations as the "only true sacrament on the Earth today," you may well gain a greater understanding of just how important they are. Such understanding of their importance makes it imperative that we learn to make good preparations. The quality of the preparations may be primarily dependent on the quality of our attitude toward them. I have often said that one of the reasons that biodynamic agriculture has done so poorly, especially in this country, but elsewhere in the world as well, is that we do not really believe in it or the preparations. In comments to Ehrenfried Pfeiffer, Rudolf Steiner stated: "the benefits of the biodynamic preparations should be made available quickly as possible to the largest possible areas of the entire Earth, for the Earth's healing." Elsewhere, and on frequent occasions, Steiner expressed the hope and even the necessity that anthroposophy should be widely established in the world by the end of the twentieth century to prevent a tragic future for human evolution. As a somewhat neglected offspring of anthroposophy, biodynamics can hardly be regarded as widely established or as affecting the largest possible areas of the Earth in the year 1994, just six years short of the end of the twentieth century. If we regard Steiner's words not as idle conversation, but as something of profound importance, you may come to the same conclusion, which I have now reached. Our belief in the biodynamic preparations empowers them, and our use of the preparations on the Earth at the end of this century is an absolute spiritual necessity. So much to do and so far to go in such a short time.

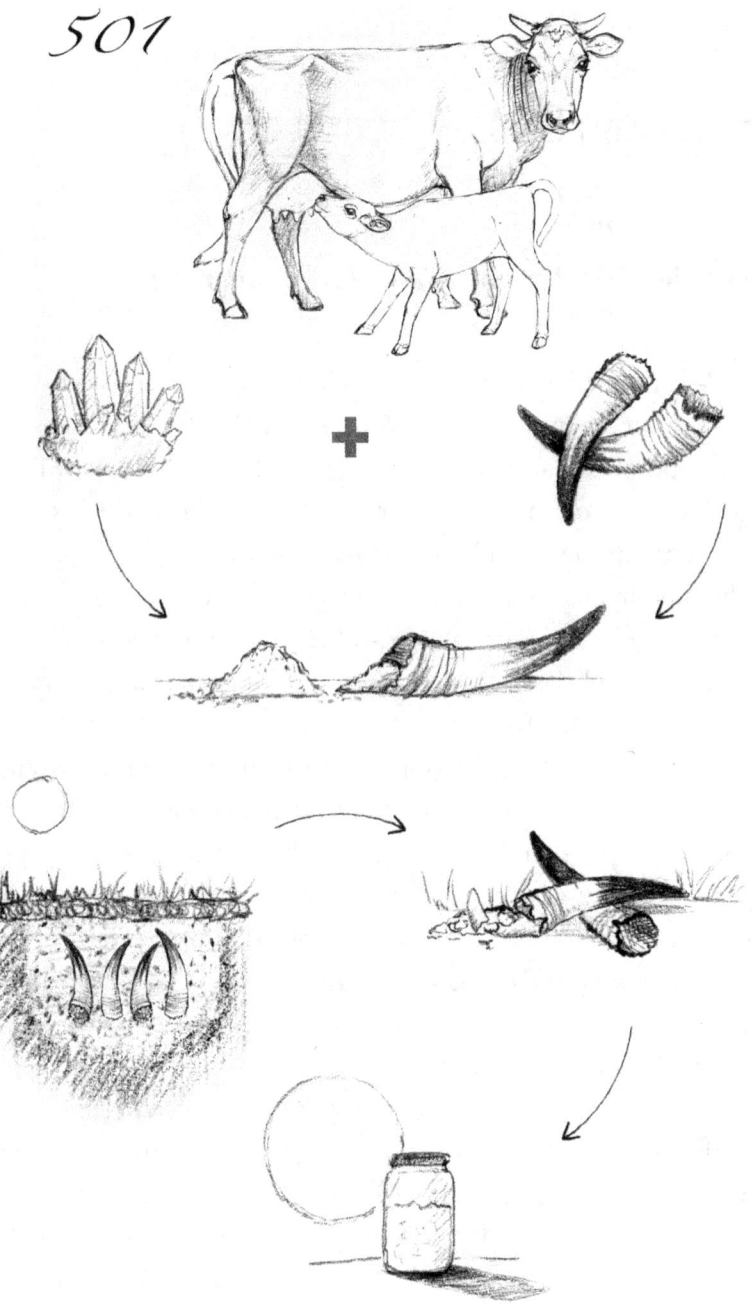

501: Horn Silica Preparation

Hugh J. Courtney
Applied Biodynamics, *no. 12, summer 1995*

Of all the biodynamic preparations, Steiner has less to say about the actual preparation horn silica, BD501, than any other with the exception of valerian, BD507. Is this because he placed substantially less value on this particular preparation? From the history of its use, or perhaps one should more properly say, its lack of use, by many of the biodynamic practitioners especially here in America, it would definitely seem to be of far less value than almost any other preparation. Besides his relatively sparse words on the subject of what is now known as BD501, Steiner presents us with a description of the materials to be used in making the horn silica preparation that to a very large extent has been ignored for virtually the entire seventy plus years of the biodynamic movement in the world.

Let us take a good hard look at exactly what Steiner has to say about the horn silica, not only in the primary description in *Agriculture: Spiritual Foundations for the Renewal of Agriculture,* lecture 4, but elsewhere throughout that course. I will also refer to the appendices and anecdotal material from the personal conversation, held with Dr. Steiner, which refer to this preparation, as reported and published with this lecture course in the 1993 edition. In addition, I will refer to some of Steiner's other works to detail further insights to be gained from his spiritual-scientific research. Having just finished his description of the "spiritual manure" now known as BD500, Steiner would have us follow up its use with another step:

501: Horn Silica Preparation

> Once again take cow horns, but this time, instead of stuffing them with manure, fill them with quartz that has been ground to a powder and mixed with water to the consistency of a very thin dough. Instead of quartz, you could also use orthoclase or feldspar. And now, instead of leaving the cow horn in the ground through the winter, let it stay underground all summer, and then take it out in late autumn. Save the contents until the next spring and then take what was exposed to the summer life of the Earth and treat it just as you did the manure.[1]

While Steiner still has a bit more to say about the horn silica preparation in this lecture, I would like to pause here to examine more carefully what it is that he has just told his Koberwitz audience up to this point about "BD501". Although his first reference to a silica-containing mineral rock source is very clearly to quartz, he also says "you could also use orthoclase, feldspar." To my knowledge, this is the statement that those in the biodynamic movement in the world have basically ignored for the entire history of the movement. In order to provide a proper basis for my contention that we must assign a great deal more importance to this seemingly offhand comment of Steiner's, I would like to examine and provide commentary on a few of the definitions that are detailed in the sidebar (opposite).

Although Steiner gives us relatively few words about the horn silica *per se*, he does have a good deal to say about silica itself, describing in one place that "by means of the silica-containing sand...life-ether and chemical-ether, as we can call them, first enter into the soil and then take effect as they stream back upward." In the very first lecture of the course, he tells us that "Plant life as we know it today can thrive only when these two forces—the forces of substances like lime and like silica—are in equilibrium and are working properly together." The substance that helps establish this equilibrium and ensures that the lime and the silica (two very disparate materials in

---

1 Steiner, *Agriculture: Spiritual Foundations for the Renewal of Agriculture*, p. 74

Definitions

SILICA: A white or colorless crystalline compound, SiO_2, occurring abundantly as quartz, sand, flint, agate, and many other minerals and used to manufacture a wide variety of materials, especially glass and concrete.

SILICON: A nonmetallic element (Si) occurring extensively in the Earth's crust in silica and silicates, having both an amorphous and a crystalline allotrope and used doped or in combination with other materials in glass, semiconducting devices, concrete, brick, refractories, pottery, and silicones. Atomic no. 14; atomic weight 28.086; melting point 1,410°C; boiling point 2,355°C; specific gravity 2.33; valence 4.

QUARTZ: 1) A very hard mineral composed of silica, SiO_2, found worldwide in many different types of rocks, including sandstone and granite. Varieties of quartz include agate, chalcedony, chert, flint, opal, and rock crystal. 2) One of the commonest minerals, silicon dioxide, SiO_2, having many varieties that differ in color, luster, etc., occurring in crystals (rock crystal, amethyst, citrine, etc.) or massive (agate, bloodstone, chalcedony, jasper, etc.), an important constituent of many rocks. It is piezoelectric and is cut into wafers used to control the frequencies of radio transmitters.

ORTHOCLASE: A variety of feldspar, essentially potassium aluminum silicate ($KA_1Si_2O_8$), characterized by monoclinic crystalline structure and found in igneous or granitic rock. Also called potash feldspar.

CLAY: 1a) A fine-grained, firm earth material that is plastic when wet and hardens when heated, consisting primarily of hydrated silicates of aluminum and widely used in making bricks, tiles, and pottery. 1b) A hardening or non-hardening material having a consistency similar to clay and used for modeling. 2) Geology – A sedimentary material with grains smaller than 0.002 mm in diameter. 3) Moist, sticky earth, mud. 4) The human body as opposed to the spirit.

LIME: 1) The oxide of calcium (CaO), a white caustic solid prepared by calcining limestone, etc., used in making mortar and cement. When treated with water, it produces calcium hydroxide, $Ca(OH)_2$, or slaked lime. 2) Any Calcium compounds for improving crops or lime-deficient soils.

Steiner's description of their nature) are indeed working together properly, is none other than clay that "mediates between them. Clay is closer to silica, but it still mediates toward lime."[2]

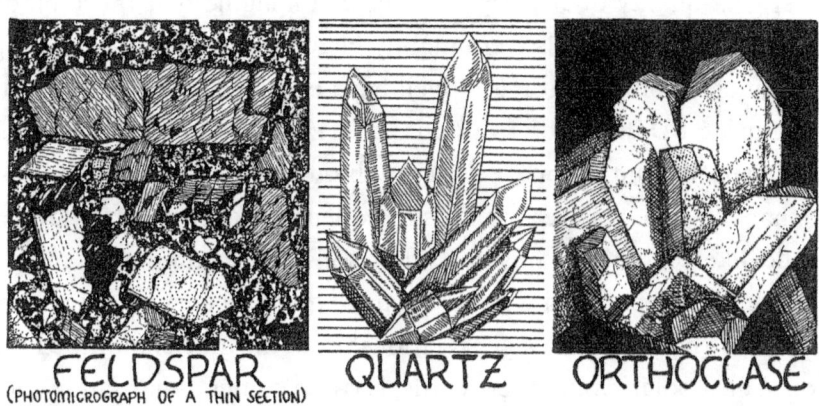

FELDSPAR (PHOTOMICROGRAPH OF A THIN SECTION) QUARTZ ORTHOCLASE

When we look at the accompanying definitions, we can readily see that clay, composed of hydrated silicates of aluminum, is indeed closer to silica, SiO_2. Orthoclase with its potassium, K, and especially a feldspar containing calcium, would have a much more direct relationship to lime and "related substances such as potash and sodium"[3] than would a pure quartz crystal containing only SiO_2. (Within the periodic table of elements, calcium [lime] is strongly related to magnesium, potassium, sodium, and perhaps also barium among others. When soils are tested more than superficially, the so-called cation exchange capacity calculated at a ratio of seventy percent calcium, ten percent magnesium, eighteen percent potassium and two percent sodium is generally regarded as the ideal.)

For years, however, the general biodynamic practice has been to use only quartz rock crystal in the making of the biodynamic preparation BD501, horn silica. Indeed, such a biodynamic authority as Lily Kolisko in the monumental work, *Agriculture of Tomorrow,* by referring only to rock crystal as the source material of BD501 would

---

2   Ibid.; also *American Heritage Dictionary of the English Language,* 3rd ed.
3   American Heritage Dictionary of the English Language, 3rd. ed.

seem to imply that one need not consider any other possibility. Alex Podolinsky, who is well known for the tremendous accomplishments made by biodynamic agriculture in Australia, would choose only the finest gem-quality quartz crystal. One can imagine that those biodynamic farms where soils were mainly heavy clay or containing a significant clay component might well have obtained impressive results while using BD501 made from pure quartz rock crystal. If such soils were predominant amongst the early biodynamic farms, it may just be that the tone was set for all subsequent production of BD501. I would like to suggest, however, that soils that contained a mostly sandy component could well have gained greater benefit from a form of BD501 that had a stronger relationship to the lime end of the limestone/silica polarity. In other words, I would contend that sandy soils should be using horn silica, BD501, made from orthoclase or feldspar with their constituents of calcium or related elements.

I shared the preceding "theoretical construct" with Dennis Klocek when he visited the Josephine Porter Institute. Dennis, with imagination and insight, took this theory to a destination well beyond the boundaries I originally conceived. With his knowledge of geology and his knowledge of plants, he tailored a form of BD501 to a particular situation and to a particular crop in a way that became the way I practiced biodynamic agriculture afterward.

I would like to think that we will begin to practice this kind of "imaginative biodynamics" much more widely in the future. Not that I would wish to encourage unbridled and unthinking experimentation along these lines merely for the sake of experimentation itself. We should, indeed, approach any new forms of the preparations, including BD501, only by thorough research and study of both the plant and the mineral kingdoms.

In my own efforts through the years to "think into the mystery of BD501," I have focused on two other aspects of the question. Neither of these aspects is from a direct statement by Steiner in the *Agriculture* lectures. Rather they come from the question-and-answer sessions in the one case and, in the other case, from the anecdotal material appended to the course. The stimulus for the first point of focus comes near the end of the second discussion where the following question and answer exchange takes place:

> "How should we grind the quartz or silica? With a small mill or with a mortar and pestle?"
>
> It's best to do it in a mortar, to begin with, and you will need an *iron pestle*. Grind it to a very fine, mealy consistency. *If you use quartz,* you will first need to grind it as far as possible with the mortar and pestle, and then grind it some more on a glass surface. It has to be a very fine powder, which is very difficult to achieve with quartz."[4]

I italicized what I regard as two key portions of Steiner's response to this particular question so as to direct your attention to them more adequately. The emphasis on the phrase: "If you use quartz" is to underscore my contention that the alternate choices of orthoclase and feldspar were definitely within Steiner's frame of reference by his use of the word *if*. It is quite possible that the nature of the rest of his answer may have been responsible for the subsequent focus by the biodynamic movement on the use of quartz rock crystal as the choice for making BD501.

The emphasis on the words *iron pestle*, however, is what has directed my experimentation along one particular avenue for possible source material to be used in making an "experimental" BD501.

I was struck from my very first reading of this statement by Steiner's choice of the word *iron* as the material for the pestle. This became even more interesting when I actually attempted to use an iron pestle for grinding quartz rock crystals. The most remarkable thing about this effort was that I was actually abrading some iron particles from the pestle virtually equal to the quantity of quartz powder I was creating in the process. From reading elsewhere in Steiner's works (in places I cannot specifically identify for this article), I was well aware

---

4   Ibid., p. 112.

of the importance Steiner placed upon iron, particularly regarding the relationship of iron to the Archangel Michael. Is it possible that one should deliberately incorporate iron within a horn silica preparation? This possibility came even more strongly to my attention when I learned that the famous "Fairy Stones" of Patrick County, Virginia, a rock crystal known as staurolite, were composed of ferrous aluminum silicate. Prompted by this recognition, several years ago I made a small batch of this "Fairy Stone" BD501. Unfortunately, there has been almost no possibility of conducting the kind of experiments this material seems to warrant, so it is merely a hope, at most, for future experimentation.

My other effort at "thinking into the horn silica mystery" was prompted by the report by Ehrenfried Pfeiffer in the section titled "Further Agricultural Indications by Dr. Steiner":

> For the silica preparation [BD501], Dr. Steiner said it would even suffice to mingle and knead up a piece of quartz of the size of a bean with soil from the land that is afterward to be sprinkled, and put this mixture into the horn. This would already contain sufficient silica radiation if a little of it was dissolved and stirred.[5]

I chose to take the quote from the Adams translation for two reasons in particular:

1. At the time concerned, it was the only translation in print.
2. The new translation on page 253 uses wording as follows: "fill the horn with a bean-sized piece of [powdered] quartz," and it seems overwhelmingly evident that "powdered" and "bean-sized" quartz are mutually exclusive terms. Steiner is referring to a *piece* of quartz, not powdered quartz.

---

5   See ibid., p. 253. In the Creeger/Gardner translation, the word *powdered* is set in square brackets, indicating an editorial addition. —ED.

501: Horn Silica Preparation

When I first began the preparations work in 1984, I was deeply concerned at the extremely modest amount of BD501 being ordered in comparison to the BD500, horn manure, and the compost preparation sets, BD502–507. In particular, I wondered if I was not making BD501 properly or whether I was not using the best source of quartz. At one point, I expressed my concern to Dr. Mary Lee Plumb-Mentjes, and she shared the problem with her husband, Conrad Mentjes. It seems that Conrad is a highly accomplished "rock hound" and felt that a particular source of quartz crystals might be ideal for making the BD501. He referred to this particular source as "rectorite," although rectorite, as it turns out, is an almost rubbery clay matrix into which quartz crystals had been embedded. They themselves were extremely small crystals, even microscopic, and of extremely fine form. They were needle-like in shape, with a very high percentage being double-terminated-altogether, a most remarkable material. Given the association with the clay matrix of "rectorite" and the previously cited indication of the "bean-sized" quartz reported by Pfeiffer, I chose to make a version of BD501 from this quartz sand material without grinding the material at all, since so much of it was already microscopic in size. The result was what has sometimes been labeled BD501(c) or "rectorite" crystal sand BD501. It is this material that we used in our initial sequential spraying efforts (*Applied Biodynamics*, no. 6), with impressive results.

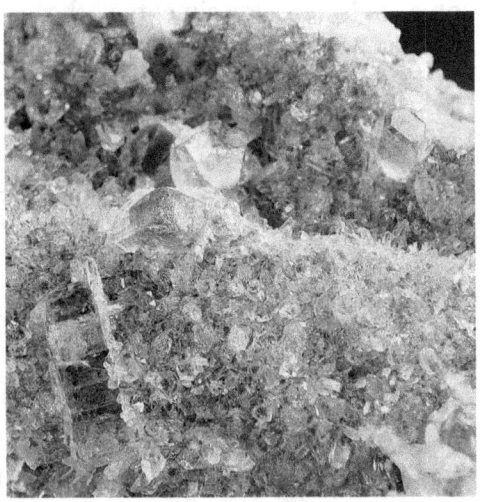

Rectorite

This is also the material I used when I obtained the first observable results from the use of BD501, which were dramatic enough to be inescapable. Those results were observed after spraying four 100-foot rows of potato plants that had a substantial infestation of the Colorado potato beetle and larvae. The following morning, after the spraying, all beetles and larvae were dead. It still remains my personal choice as the form of BD501 for application here in Woolwine, Virginia. I should add that the crystal sand has a markedly reddish color, an indication of high iron content. One still has to conduct some trials and experiments with this form of horn silica in order to take it beyond the point of personal preference.

I should mention that my concern about the efficacy of the BD501 I was making was unfounded. It seems that a great many people ordered BD501 one year and not the next because they never got around to using it. They were simply "too busy in the garden and there just wasn't enough time." One certainly cannot observe the effects of a preparation, however subtle those effects may be, if one is, in fact, not even using the preparation.

My main hope in writing this article is to stimulate the biodynamic practitioner to a greater understanding of the many possibilities available in our use of BD501, so that it will become more widely used and that it will be much more regarded as a necessary part of one's biodynamic practice rather than something that gets done only "if there's time."

Let us examine the "necessary" aspect of using BD501 first of all, by turning again to Steiner's initial description. Having told his audience to treat the horn silica just as the horn manure had been treated, he then tells us about the differences in dealing with this preparation. Steiner states:

> In this case, however, you will need much smaller quantities; you can take a portion the size of a pea, or maybe no bigger than a pinhead, and stir it into a whole bucket of water. This, too, needs to be stirred for an hour. If you apply this as a fine

501: Horn Silica Preparation

spray on the plants themselves—especially on vegetables and things like that—you will soon notice how its effect complements and supports the influence coming from the other side, from the soil, as a result of the cow-horn manure.[6]

Two particular phrases in the preceding quote stand out in my mind: the "fine spray" used for the BD501 and the fact that "its effect complements and supports" the BD500. Earlier, Steiner has expressed the need for the BD500 to really unite with the soil. In calling for a "fine spray," is he not also describing a process that will really unite the horn silica with the air or the atmosphere around the plant? Through the stirred moisture carrying these two preparations, one in the soil and one in the air, an entirely new growing zone has been created for the plant in that all-important "diaphragm"[7] area Steiner refers to in lecture 2. To concentrate one's biodynamic effort in only the area below the diaphragm, using only BD500 or BD500 plus the compost preparations (BD502–507), will lead to a plant that only partially expresses its full biodynamic potential.

BD501 is, indeed, required to support and complement BD500. Precisely how these two preparations manage to be complementary is a question that calls for much study, and we will hope to share additional thoughts on this subject at some time in the future.

It is very often the case that a person will finally get around to using BD501 and, because no dramatic and unmistakable differences can be observed in the plants, an assumption is made that it was a waste of time to use it. This is perhaps because we have not looked closely enough at the plants being affected by the silica.

In speaking of curative treatment with siliceous substances in human medicine, Steiner speaks of the "peripheral operation" of the silica and its "formative activity," plus "its tendency to harmonize deformations," which he further identifies as "deformations that

---

6   Ibid., p. 74.
7   Ibid., p. 28.

remain in the physiological realm." If the physical plant is strengthened in its form, this may not be so obvious, since a tomato plant, for instance, still maintains its same general appearance after spraying with BD501 as it had beforehand. Closer observation, however, will reveal a crisper, more sharply defined plant that takes on a more crystalline appearance. If we have not observed any "deformations" in the plant because they have not yet assumed a noticeable manifestation, we will not be likely to observe that the plant takes on a more "harmonious" appearance. With BD501, we are dealing with a more subtle level in its effects than with BD500 where changes in the soil can be more easily noticed. At times, however, one can experience the forces connected with BD501 in a dramatic fashion. Rick Walker in Mooresville, North Carolina, tells of spraying BD501 on a row of Red Sails lettuce, which still had a totally green color when he began spraying. While still spraying the row he looked back at the previously sprayed lettuces and saw that they were all taking on the desired red color even as he watched. It is not too surprising then, that Rick insists that a crop is not yet "biodynamic" if it has not been treated with BD501.

In the future, it will be very important to examine other aspects of the horn silica and how it expresses its effects within agriculture. In particular, I hope to address the question of how geographical and seasonal forces may enter considerations of how to use BD501.

502

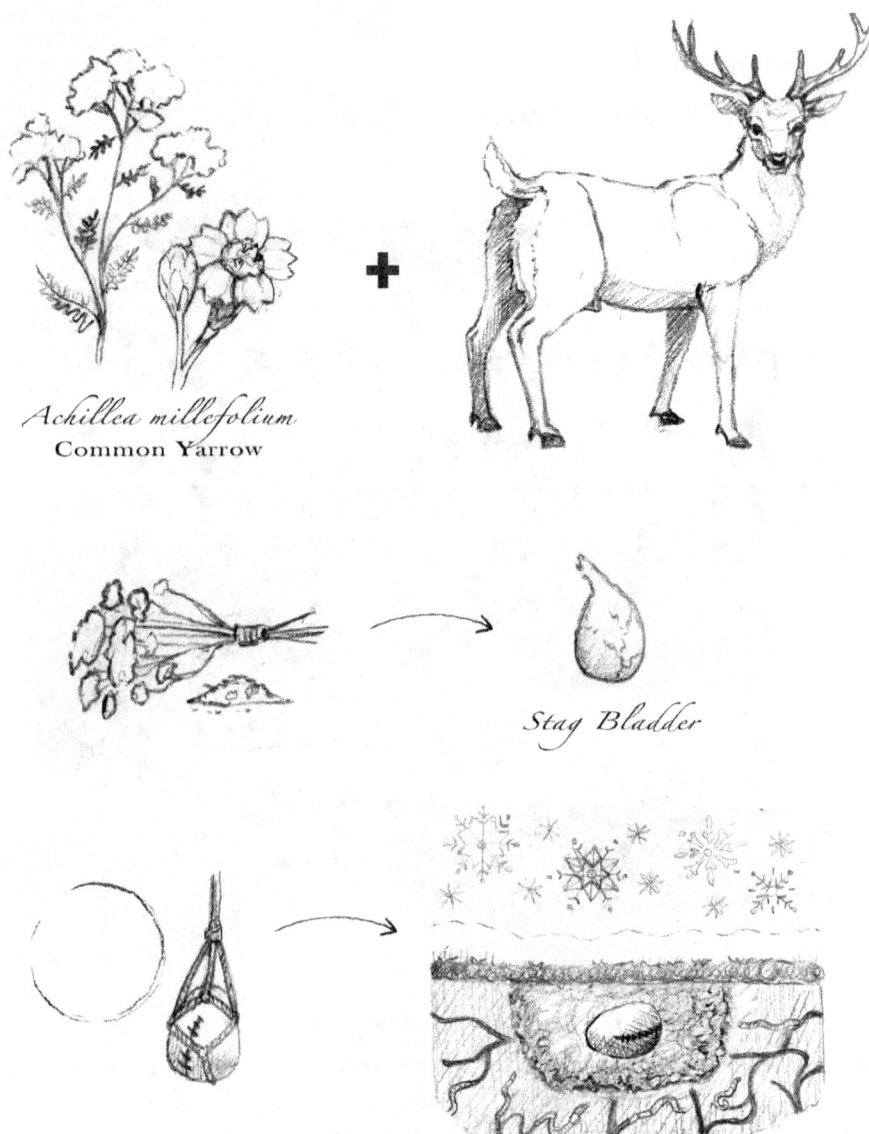

Achillea millefolium
Common Yarrow

Stag Bladder

502: How to Make the Yarrow Preparation

Patricia Smith
Applied Biodynamics, *no. 37, 2002.*

"*Yarrow is indeed a miraculous creation. No doubt every plant is so; but if you afterward look at any other plant, you will take it to heart all the more, what a marvel this yarrow is.*"
(Rudolf Steiner, *Agriculture Course,*
trans. G. Adams, p. 91)

502: How to Make the Yarrow Preparation

These are the step-by-step procedures we follow to make the yarrow preparation at the Josephine Porter Institute in Woolwine, Virginia. Yarrow (*Achillea millefolium*) is located in fields, meadows, and roadsides, growing one to three feet high, with finely dissected leaves and disks of whitish blossoms on sturdy, stiff stems.

Time of harvest: Late spring to early summer, after the flowers have bloomed for some time. Flower heads have a relatively long time from initial bloom to fully ripe seed formation.

Achillea millefolium *is known by its more common name, yarrow. Other common names include: noble yarrow, milfoil, nosebleed, sanguinary, soldier's woundwort and thousandleaf. (photos © JPI staff)*

Blossoms to harvest: Look for the flower heads that begin to protrude, losing their flat, disc-like appearance, and the petals starting to turn downward. Harvest yarrow using clippers, or a sharp knife or scissors. Make sure you take a generous amount of stem, at least three inches. This is for ease of handling when later clipping the tiny florets from the stem. Yarrow flowers do not contain a great deal of moisture so you can cut them right away for use in the BD502 preparation, or leave them out on racks, screens, or an open area to dry further for a few days or weeks. Once dried, they can be stored in paper grocery bags for up to several years, but it is best to use them within a year. Fresh yarrow for making BD502 is the ideal situation if you have enough supply.

Using small hand scissors, a sharp-bladed pocket knife, clippers, or even a sturdy thumbnail, cut the individual flowers, removing them from the umbel structure of stems. Our preference at JPI is to have virtually stem-free florets to use in stuffing the bladders. This is most important because the stiff coarse stems, if left attached when being stuffed into the sheath, can puncture the walls of the stag bladder. Reserve the stems and leaves for later use in making a "tea" for moistening the blossoms and softening the stag bladders before stuffing.

Make sure to cut more than enough blossoms. Lots and lots of yarrow blossoms are required to fill even a small stag bladder. When you have finished clipping, the bowl of tiny florets will vaguely resemble a bowl of rice or other grain.

Left: These yarrow blossoms with flower heads protruding and petals turning downward, are ready to be harvested. Right: Make sure a sufficient supply of yarrow blossoms have been clipped before starting to stuff the stag bladder. (photos © JPI staff)

The Stag Bladder

Unless you are a deer hunter, or know someone who is, procuring a stag bladder can be a challenge, but not impossible. Check with hunters in your locality to see if a stag bladder could be saved from the local deer or elk population harvest in the fall. It may even be possible to obtain the cooperation of a local game warden who will

502: How to Make the Yarrow Preparation

allow you to harvest bladders from fresh roadkill. If you find someone willing to help, ask if they will reserve the bladder with the testes and male organ intact to ensure that the bladder is from a stag. Also, if possible, ask the hunter to refrain from rinsing the bladder out, leaving the urine within. This urine can be reserved and stored in a glass container for later use to rinse the bladder before stuffing it with yarrow. Once you have a bladder you can choose to work with it fresh or freeze it for future use. If you choose to freeze it, make sure water covers the entire bladder. Do not attempt to vacuum seal the bladder; it becomes tough from lack of moisture and is difficult to use.

Cut off the stems completely from the individual flowers to prevent the bladder from being punctured. (photo © Patricia Smith)

Preparing the Stag Bladder for Use as a Sheath

Thaw the frozen stag bladder in water. Once thawed, place the bladder on a small cutting board. Smooth out the bladder and testes being careful not to puncture or cut into the tissue. With a knife, remove part of the male organ, leaving about one inch, exposing the urinary tract in the center.

Using a small air pump with a needle attachment, insert this deep enough into the urinary tract opening so that the end of the needle enters the bladder itself. Cut a five-inch piece of yarn, tie and secure a single knot around the tissue that surrounds the needle attachment.

At this point, it works best as a two-person operation. Although it is possible for one person to do this, it involves some difficulty. As the first person keeps the rather slippery tissue tightly over the needle with the yarn tied loosely around the opening of the bladder but behind the testes, the second person uses the small air pump to inflate the bladder. If you encounter a bladder that has a puncture or air leak, all is not lost. See the sidebar below if this situation arises.

Depending upon the size of the bladder, it can expand from two to four inches across. Once the bladder is expanded into a sizable "balloon," the second person is now ready to tie off the opening. The first person holds the bladder in place, being careful not to let any of the air escape while the second person ties the yarn into a secure knot at the opening. The first person removes the needle from the bladder as the knot in the yarn is tightened down.

Left: With an air pump, inflate bladder and use fingers to secure it over the needle attachment. Right: An inflated stag bladder dries with the testes intact. It usually takes twelve hours for the sheath to dry before stuffing with yarrow blossoms. (photos © Patricia Smith)

HOW TO INFLATE A PUNCTURED DEER BLADDER

Insert a wooden, unsharpened pencil that has been lightly coated with vegetable oil into the opening of an un-inflated balloon. Slide the balloon into the bladder through the urinary tract opening. Remove the pencil and insert the needle attachment of an air pump into the neck of the balloon. Inflate as directed in "Preparing the Stag Bladder for Use as a Sheath." These punctured bladders are mainly reserved for use in repairing the stuffed stag bladders in the fall. However, if no other bladders are available, the balloon-inflated bladders can also serve as a suitable sheath for the yarrow preparation.

Drying the Inflated Stag Bladder

Once the knot is securely in place, an additional piece of string about ten inches long is tied around the opening. Tie and suspend the inflated bladder from a pole or line and leave to dry thoroughly for perhaps twelve hours or so, depending upon temperature and humidity. If drying several bladders, make sure they are not touching each other as they can easily become glued together, and damage to a bladder may occur when separating them.

After drying, if the bladder is not going to be used right away, store it loosely in a sealable plastic or paper grocery bag in the freezer. In this way, it is kept fresh and away from animals and insects, particularly flies, which seem to prefer the bladder as a prime site for laying eggs that quickly hatch into voracious larvae. Due to its fragile nature, make sure nothing is placed on top of the inflated bladder. It is best to use the deer bladder within one year, but it can be stored in the freezer for up to three years. Time of year to make the yarrow preparation (stuffing the stag bladder): Late June to mid-July.

Preparing to Stuff the Stag Bladder

To make the tea for the BD502 preparation, simmer water in a large saucepan or pot, and add several handfuls of yarrow stems and leaves. Simmer for approximately twenty minutes.

In a bowl filled with clipped yarrow blossoms, add the simmered yarrow tea, to which a small amount of cool water has been added to avoid scalding the blossoms. Sufficient liquid should be used to lightly moisten the blossoms, avoiding making them soggy or dripping wet. A good rule of thumb is to squeeze a handful of the moistened yarrow blossoms; if a great deal of water can be squeezed out, it is too wet; if only a drop or two, it is just right. Whether freshly dried or stored in the freezer dried, the bladder needs to be rehydrated to become soft and pliable enough for stuffing. To soften, place the dried deer bladder in a small bowl of warm water blended with yarrow tea.

Since the bladder is like a balloon and quite buoyant, turn it several times in the warm liquid. It should be pliable in about five to ten minutes. At this time the interior of the bladder can be rinsed with any reserved deer urine.

Stuffing the Stag Bladder with Yarrow

After the bladder is softened, make a vertical one to two-inch slit with scissors near the site of the knot. Slide a sausage-stuffing funnel that has had an inch cut off the narrow end into the opening of the bladder and start to feed the moistened yarrow blossoms through. A wooden dowel rod is helpful for this purpose to push the moistened yarrow along. Once the sheath is about half full, start to firmly press down the blossoms with the use of the dowel rod or thumb and fingers. This is important because if the sheath is loosely packed it will contain too much air, causing the yarrow blossoms to dry out, thereby halting the transformation process.

When the sheath is nearly filled, remove the funnel and stuff the rest by hand. You know you are finished stuffing when you can

502: How to Make the Yarrow Preparation

press your finger on the outside of the sheath and it will not make a depression. The sheath should be a firm consistency like that of a softball. While some European BD practitioners stuff bladders quite loosely and with substantial stem material attached to the flower, the practice at JPI is to stuff as tightly as possible. You probably will be amazed at how many blossoms it takes to fill the sheath to make the yarrow preparation. Once filled, the sheath opening is ready to be sewn up.

Left: Soak the inflated stag bladder in warm water blended with yarrow tea for about five to ten minutes to soften before stuffing. (photo © JPI Staff) Right: A little further down from the area that was tied off, pierce the bladder with scissors and then cut a one-inch opening. This will allow enough space for the moistened yarrow blossoms to be filled in whether using fingers or a funnel. (photo © Patricia Smith)

Using heavy-duty white cotton thread and a large needle, double thread the needle and sew up the opening using a hem-, blanket- or whip-stitch, or whatever stitch you know that can do the job. Squeeze the two sides of the opening together as you stitch to complete the closure.

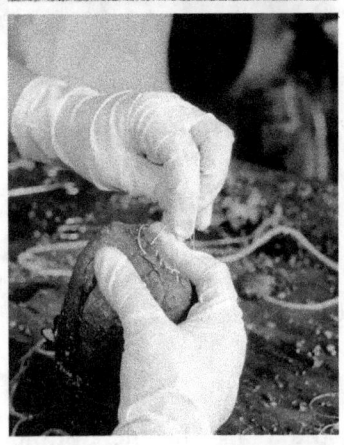

The sewing completed, the yarrow preparation receives a protective covering of cheesecloth before being hung outside. Fully cover the bladder with the cheesecloth as the covering is necessary to shield the bladder from splitting and cracking in the summer sun. It may provide protection from birds and insects, as well, although we have never experienced any damage from birds here in Woolwine, Virginia. You can purchase cheesecloth at the hardware or sewing store in a package of about two square yards. To form a cover, lay out a piece of cheesecloth, a layer thick. This should measure about eighteen inches square and should be sufficient to cover most bladders. Place the yarrow preparation on the cheesecloth.

When cutting the cheesecloth, allow extra material for the entire sheath to be covered and for tying knots. Bring both sides of the cheesecloth together and make a double knot directly over the top of the yarrow-filled bladder to fit snugly, so it won't fall out. With the extra cheesecloth (you should have at

Top: Once the stuffing is nearly completed, remove the funnel and fill the rest by hand. (photo © JPI staff) Middle: After fully stuffing, pinch the sides of the bladder together to secure a tight stitch. Bottom: Sew up the stag bladder using a blanket stitch or a similar reinforcing stitch. Some practitioners prefer to wear latex gloves since handling the bladder can leave a distinct odor on hands for a few hours. (photos © Patricia Smith)

least four to six inches left on each side), make another double knot at the top to form at least a four-inch loop.

Summer Exposure

The covered yarrow preparation is ready to hang outside. Find a location high enough from the ground that animals will not be able to reach it. Also, make sure there is enough space to hang it away from a wall, tree branch, or tree trunk so that the yarrow preparation will not crash against it during high winds. Once you have chosen an appropriate location, secure it to a pole, hook, or branch so it cannot fall or blow away. The yarrow preparation hangs outdoors from early summer to fall.

Ready for Fall

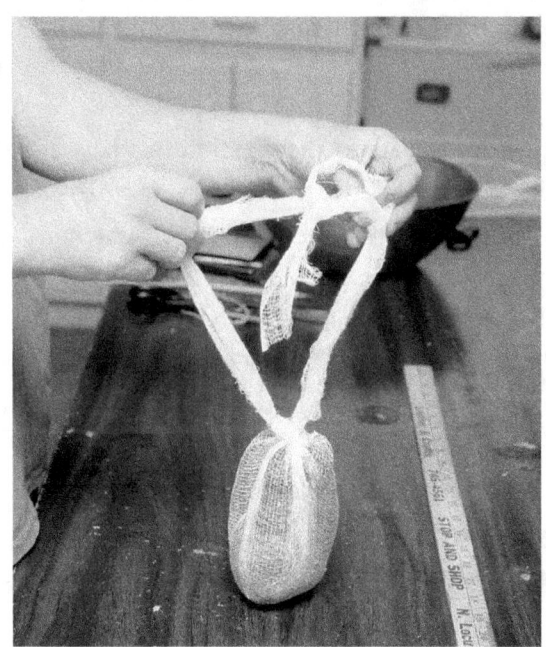

Left: Finish securing the second and final knot to fit snugly so the yarrow filled bladder will not fall out of the cheesecloth (photo © Patricia Smith) Right: A From early summer to fall, the yarrow preparation hangs outdoors (photo © JPI staff)

After the fall equinox, in late September or October, take down the yarrow preparation and remove the cheesecloth. Examine the sheath for any rips or tears. If there is a rip, you can use a spare stag bladder (balloon-inflated) saved for such repairs. Soften the spare bladder in warm water. Once softened, make a small vertical slit to remove the balloon carefully; it should come out of the bladder pretty easily. Once removed, make the vertical slit longer to fit over/around the yarrow preparation in need of a repair. It should fit as snugly as possible. Sometimes this is not possible and the best fit with the spare materials available can look wrinkled and baggy over the yarrow preparation. This is acceptable since the main objective is to cover the tear. You can further smooth out the wrinkles by wetting your fingers with water and applying this to the loose-fitting bladder. If there isn't a spare stag bladder for repairs, simply sew up the torn area of the sheath with a gentle stitch such as a blanket stitch or hemstitch.

Burying the Yarrow Preparation

The yarrow preparation is ready to go into the ground, but here at JPI we prefer that it not go in "as is." For ease of recovery and protection, it must be enclosed with fine mesh copper screening, which can be found at hardware stores. Fiberglass screening is an alternative, but never use hardware cloth (galvanized steel) as the Mars/iron relationship is opposite in polarity to the Venus/copper relationship. Once the yarrow preparation is enclosed in the screening material, fold the screen over at the edges and staple about every three inches along the outer edge. Without the screen, the yarrow preparation can be eaten up by underground critters such as insects and moles, or even be of interest to foxes and dogs. Locate the pit in an area with good drainage and far enough away from tree roots. To prepare the BD502 pit, dig to a depth of one to two feet deep and at least one to two feet wide. Bury the yarrow preparation in the late afternoon around dusk. Bricks may also be used to cover the bladders and to protect them from a shovel wielded too enthusiastically. Mark the area with

502: How to Make the Yarrow Preparation

a large metal or wooden stake. This is very important because, without a marker, it becomes quite difficult to find later. After burying the yarrow preparation, one might choose to meditate and pray on the purpose of this preparation before leaving the site.

Digging up the Yarrow Preparation

The yarrow preparation stays in the ground through the remainder of the fall, winter, spring, and first few days of summer. After St. John's Day (June), the yarrow preparation should be finished and ready to dig out of the pit. Some practitioners choose to dig up the yarrow as early as Easter.

However, since this is substantially less than the year that Rudolf Steiner indicated for the yarrow to spend in the bladder, here at JPI we focus on St. John's Day for digging them up.

Locate the area where you buried the BD502. With a shovel, dig the soil out of the pit, being careful not to dig into the yarrow preparation. When you start to see the copper screen or protective bricks, stop digging and start brushing away the soil by hand.

In the summer, carefully dig up the yarrow stuffed bladders enclosed in copper screening. (photo © JPI staff)

Gently pull the screen out of the pit, keeping it horizontal so none of the contents roll around or break apart, and place the yarrow-filled bladders into a shallow tray. Remove the staples from the copper screen and carefully remove the finished yarrow preparation. The bladder should be a visibly thinner membrane and the yarrow should be completely transformed: light brown in color and an almost feathery light texture. Peel away the sheath, gently scraping the yarrow clinging to the sides with a pocket or butter knife or similar instrument. Screen the yarrow preparation through a quarter-inch mesh screen.

Storage

After screening it, the finished yarrow preparation needs to be properly stored. Conditions are very important. Although naturally low in moisture, the yarrow preparation must not dry out. Store in a glazed ceramic crock, glass, or unglazed clay vessel preferably underground or in cool, dark conditions surrounding the crock with peat moss. Also, allow for the preparation to "breathe," using a loose-fitting lid such as natural stone, wood, or glass. See JPI's information sheet, "Storing Biodynamic Preparations" for further details.

Uses

Although the yarrow prep is principally used for inserting into a compost pile, it is also used as a seed soak for rye and other grasses. Read *"Achillea millefolium* Esoterica" for more information on the yarrow preparation.

Patricia Smith was the editor and contributing writer for Applied Biodynamics *from 2000 to 2010. She lives in southwest Virginia and came to the journal as a volunteer during the summer of 1997. Later that year, she was hired as an assistant preparations maker.*

Achillea millefolium Esoterica

Hugh J. Courtney
Applied Biodynamics, *no. 37, 2002*

If we are to achieve Rudolf Steiner's intentions as far as biodynamic agriculture is concerned, it is not enough that we know how to make the yarrow preparation, and occasionally use it to make compost.[1] We must exert an effort to grasp what it is that lives in this preparation and how it exerts its influence. In other words, we must do our utmost to understand the preparations, "not as substances, but as forces."

Let us revisit what Steiner says about the compost preparations as a totality. In speaking of further treatment of the manure, he says, "And the point is not merely to add substances to it, with the idea that it needs such and such substances so as to give them to the plants. No, the point is that we should add living forces to it."[2] To approach agriculture from the customary chemical or organic point of view of building our soils by supplying all the substances deemed necessary is still insufficient, and as Steiner continues, "it would still be of no use for plant-growth, unless by a proper manuring process we endowed the plant itself with the power to receive into its body the influences that the soil contains."[3] Note here that it is the imponderable quality of 'influences' that is being received from the soil by the plants, rather than merely having the plant be able "to absorb what the soil offered," which tends to lend itself to a characteristically materialistic or substance point of view.[4] In Steiner's words, the purpose of the

---

1 Steiner, *Agriculture* (UK ed.), p. 7.
2 Ibid., p. 90.
3 Ibid.
4 Steiner, *Agriculture*, p. 93.

compost preparations is to give "manure the right degree of vitality and the right consistency."[5] By such a gift, according to Steiner, we "enable it [manure] to retain of its own accord the proper amount of nitrogen and other substances that it needs in order to bring vitality to the soil."[6] Once again, note that we are bringing another imponderable, "vitality," into our considerations rather than the material substance of the chemical or organic approach.

At the point when Steiner begins to introduce the first of the six compost preparations, yarrow (*Achillea millefolium*), he expresses the concern that "we need to make sure that the major elements of the organic realm—carbon, hydrogen, oxygen, nitrogen, and sulfur—can come together in the right way with other substances, with potash salts, for instance."[7] Steiner then goes on to describe the role of potash salts in the following words:

> It is generally known that plants require a certain amount of potash salts for growth, and that potash salts, or potash in general, tends to restrict the growth process to those parts of the plant that usually become the solid framework, that is, to the stems and stem-like parts.[8]

This last statement is the general agricultural chemical and organic view of the importance of potassium for plant growth. However, Steiner then makes another statement that takes us beyond this "structural" or material view of the role of potash. He goes on to say:

> But in the context of the interaction between the plant and the soil, it is important to transform this potash so that it relates itself properly to the forming of the proteinaceous material that constitutes the actual body of the plant.[9]

---

5   Ibid.
6   Ibid.
7   Ibid., p. 94.
8   Ibid.
9   Ibid.

Here we have a very important distinction between the "solid framework" of the plant and the "proteinaceous material" that forms the body of the plant. In the solid framework, we can recognize cellulose (see sidebar) as one major product due to the presence of potash salts. Since potassium is not a chemical component of cellulose, we may assume that it at least serves a role as a catalyst in cellulose production. However, when Steiner speaks of the role of potash in the formation of proteinaceous material, he points to the importance of "transforming" the potash so that it can play a role in forming protein. In essence, potassium is still playing the role of a catalyst but at a different level of activity. A picture that comes to mind in digesting this contrast between cellulose and protein and the possible role played by potassium in both cases, is that in cellulose production, potassium is acting more in the earthly (material substance) realm, whereas in protein formation, potassium is functioning at a cosmic (living/spiritual) level. It may be that the yarrow preparation serves to enhance potassium's ability to function more effectively at this second level. In this distinction between two possible levels of functioning, we begin to have a glimpse of the potassium "process" as opposed to the mere chemical substance of potassium or potash salts.

Cellulose—a complex carbohydrate $(C_6H_{10}O_5)_n$ composed of glucose units, forms the main constituent of cell walls in most plants.

There is another chemical element that Steiner speaks of when introducing us to this first of the six compost preparations. He describes yarrow as "an ideal model in bringing sulfur into relationship with the other plant substances. You might say that in yarrow as in no other plant, the nature spirits reach the height of perfection in their use of sulfur."[10] It is noteworthy that this is the only reference to nature spirits made by Steiner anywhere within the eight lectures of the agriculture course. Perhaps because of this perfection, yarrow

---

10 Ibid.

achieves the ability as Steiner says, "... to correct any weakness of the astral body" within the animal or human organism.[11] In the plant kingdom, the yarrow preparation works through the manure or compost being applied to the soil in such a way that it can "enliven the soil so that it can absorb and retain extremely fine doses of silicic acid and lead and so on that come toward the Earth."[12] Here we have substances being absorbed whose origin is from the cosmos beyond the earth. This absorption of such substances through the "enlivening and refreshing" yarrow preparation enables the manure as Steiner says, "to counteract the unavoidable exploitation of the land that comes about through raising crops."[13] It may well be that these extremely fine doses of substances of cosmic origin are unable to retain the cosmic forces once they impact earthly soil unless the energies of the yarrow preparation are there to aid them in retaining their cosmic connection.

1. **Protein**—any of a group of complex organic macromolecules that contain carbon, hydrogen, oxygen, nitrogen, and usually sulfur and are composed of one or more chains of amino acids. Proteins are fundamental components of all living cells and include many substances, such as enzymes, hormones, and antibodies, that are necessary for the proper functioning of an organism. They are essential in the diet of animals for the growth and repair of tissue and can be obtained from foods such as meat, fish, eggs, milk, and legumes.
2. **Amino acids**—an organic compound containing both an amino group (NH_2)
3. **A carboxylic group** (-COOH), especially any of the twenty compounds that have the basic formula $NH\ CH_2\text{-COOH}$, and that link together by peptide bonds to form protein

---

[11] Ibid., p. 95.
[12] Ibid.
[13] Ibid.

There is much more to be said about the deeper meaning behind this preparation and its stag bladder sheath, as well as the difference between cow horns and antlers. Among those who have tried to fathom this deeper meaning are authors of the following titles: *Agriculture of Tomorrow* by Eugen and Lily Kolisko identifies the connection between yarrow and the planet Venus among other things; *A New Zoology* by Herman Popplebaum examines the deer family; "The Working of the Planets and the Life Processes in Man and Earth" by C. B. J. Lievegoed, again, correlates the various preparations with planetary and other energies; *In Partnership with Nature* by Jochen Bockemühl is a botanist's view of the preparation plants; and *Earth and Man* by Karl König examines the preparations in the light of embryology, as well as from other useful directions. Regrettably, none of these titles were in print at the time of this publication.

503

Matricaria chamomilla
German Chamomile

bovine intestine

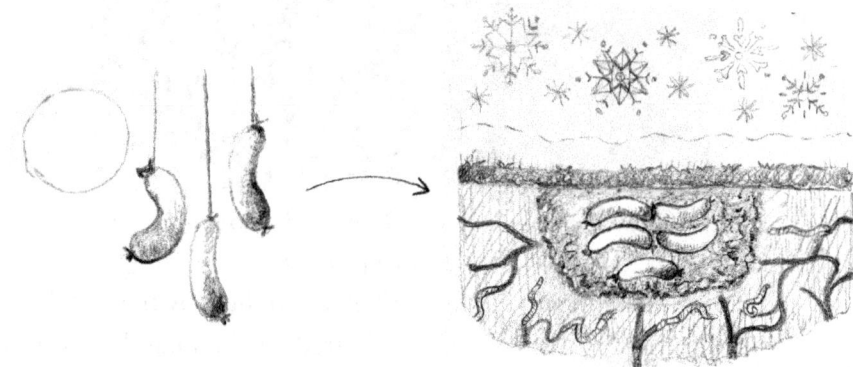

503: Chamomile:
The Healer for People and Plants

Abigail Porter with Hugh J. Courtney
Applied Biodynamics, *no. 78, 2012*

Chamomile has been used as a healing plant since ancient times. The first record of its medicinal use dates to 1550 BC in Egypt where it was associated with the Sun and the god Ra because it was valued above all other herbs for its healing qualities. The Romans and Greeks used it as well. The name chamomile comes from the Greek word meaning "earth-apple" and refers to the apple-like fragrance of the plant.

Roman chamomile (left); German chamomile (right)
(all photos by Abigail Porter)

For the Celts, it was a sacred herb given to them by the god Woden. During the Middle Ages, it was an ingredient in some love potions due to its sedative and relaxing properties. Today, chamomile is one of the most widely used herbs, and is used most frequently for digestive disorders, and to promote relaxation and sleep.

503: Chamomile: The Healer for People and Plants

In the book by Beatrix Potter, even Peter Rabbit was put to bed with a cup of chamomile tea after eating too much in Mr. McGregor's garden. It is good for "head-gut" problems, easing anxiety, which can cause stomach ulcers and intestinal distress, as well as alleviating headaches caused by digestive disturbances. It is used for fevers, pain, inflammations of all sorts, as well as for melancholy. Skin irritations and wounds are also helped by topical applications of the tea. Chamomile is antibacterial, antifungal, anti-inflammatory, antispasmodic, antiulcer, anti-viral, and has sedative effects.

In homeopathic medicine, chamomilla is used to treat patients who are peevish, impatient, angry, intolerant, excitable, whining, changeable, and restless. It is frequently used as a children's remedy for colic, teething, and insomnia. Plants benefit from chamomile as well. It is known as the "plant doctor" because it improves the health of plants growing near it. Chamomile improves the flavor of cabbage, onions, and cucumber, and increases the strength of aromatics that are grown for their use in essential oils. If planted next to an ailing plant, it is said to revive it. Chamomile tea sprayed on seedlings can prevent damping off. The tea can also be used as plant food and as a tonic for disease.

The two most common species of chamomile are *Matricaria chamomilla* or German chamomile and *Chamaemelum nobile* or Roman chamomile (previously classified as *Anthemis nobilis*). Both species have been used throughout history to treat the same ailments. About one-third of their chemical constituents are the same and the plants are similar in appearance.

Roman chamomile is a low-growing perennial. Its leaves are thicker and closer together, and the smell and taste are sharper than the German. The flowers look like small daisies with the white petals flat or angled slightly upward. The yellow cone is solid inside.

German chamomile is an annual, which self-seeds and can grow two feet tall. The leaves are spread out, airier than the Roman, and have a pleasant smell. The blossoms are usually smaller with the white petals tilting down and away from the yellow cone, which is hollow inside. German chamomile is most widely available as dried blossoms and is the most researched of the two for its medicinal properties. It is also the species that Rudolf Steiner chose for one of the biodynamic compost preparations.

Making the Chamomile Preparation at JPI

AP: Hugh, please share with us the process of how the chamomile compost preparation (BD503) is made at JPI.

HC: First, one needs to source the materials required: chamomile blossoms and bovine intestines.

Harvest and Preparation of the Blossoms

HC: It is important that you have the right variety of chamomile, which is referred to as *Chamomilla officinalis*, *Matricaria chamomilla*, *Matricaria recutita*, or German chamomile. German chamomile is the term most commonly used in biodynamic circles. In the discussion following lecture six in the Agriculture Course, in answer to questions regarding which chamomile to use, Steiner responds that it is the one that grows wild alongside the road or railroad tracks with the white petals turning downward away from the yellow cone center. He differentiates this from the variety commonly grown in gardens at that time—most likely Roman chamomile, which he said is useless for this purpose.

A sure way to tell if you have the right variety is to slice the yellow cone in half vertically. The variety to use will have a hollow center. The other varieties will have a solid center. Only the flower head is picked. The blossoms are then dried on a non-metallic screen in a warm airy place away from direct sunlight. After they are dried, they are stored in a cool place and vacuum-sealed, frozen, or in some way protected, to avoid insect damage. When picking them in the wild, one would want to make sure that no chemicals have been applied to the land or close by and that the plants have not been exposed to exhaust fumes from traffic.

503: Chamomile: The Healer for People and Plants

*Crosscut of German chamomile (left); Roman chamomile (right)
Note the hollow core of the German chamomile blossom.*

> "The preparation No. 503 is of special interest because it is made from chamomile. This plant contains a growth hormone which is a particular stimulant for the growth of yeast. The remarkable thing about this growth hormone is that it works in very high dilutions... chamomile juice is active at its best in a dilution of 1 to 8 million."
>
> —Ehrenfried Pfeiffer[1]

We do not have enough chamomile growing here on the farm to supply the needs of JPI—thirty to fifty pounds of dried blossoms annually. In addition to blossoms generously sent to us by biodynamic practitioners around the country, we order the blossoms from reliable herb companies, making sure that they are human food-grade, organic or wild harvested, and definitely not irradiated, as most imported herbs are these days. For an individual or a small farm, a couple of boxes of bulk tea would suffice. Sometimes health food stores sell bulk herbs, which would be a more cost-effective way to purchase the blossoms if one did not need very much. A whole pound of dried blossoms would be enough to stuff around six or more nine-inch-long pieces of intestines, depending on how tightly they are stuffed. That would make over one hundred units of preparation, plenty to treat manure on a larger farm [one unit treats ten tons or more of manure].

---

1 Pfeiffer, *Pfeiffer's Introduction to Biodynamics*, pp. 30–31.

> Referring to why intestines were chosen as sheath: "Now you must trace, for example, the process which chamomile undergoes in the human and animal organism, when taken as food or medicine. The bladder is comparatively unimportant for what the chamomile must undergo in the human or animal organism. In this case, the substance of the intestinal walls is far more important…" —RUDOLF STEINER[2]

At JPI, we like to produce not less than four to five gallons of finished preparation, which takes about thirty pounds of dried blossoms. One cow will supply around 130 to 140 feet of small intestines, enough for this amount, depending of course on how tightly the intestines are stuffed. The dried blossoms must be moistened before they are stuffed into the intestines. This is done by taking a few of the blossoms and making chamomile tea to moisten them. Wait until the tea has cooled down a bit and is no longer scalding before pouring it onto the dried blossoms that you have placed in a large bowl. It can be hot but not so hot that it would burn you or take the life out of the blossoms. You want them moist enough so that they stuff easily into the intestines, somewhat the consistency of thick oatmeal but not so wet that you can squeeze any water out. If they are too dry, stem ends may puncture the intestines. And if there were any long stems, you would want to cut them off before you moistened them. One to two cups of tea would moisten a pound of blossoms. You want to add the tea slowly and incrementally as you blend. If you get them too wet, just add some more dried blossoms until you reach the right consistency.

Harvesting the Intestines

AP: What is the procedure for harvesting the intestines once the cow is sacrificed on the farm?

HC: Before we separate the innards from the cow, we tie off the rectum and the esophagus so that the digestive contents and manure do not spill out over everything. Rather than cutting the abdominal sack down the middle and letting the contents fall

---

2   Steiner, *Agriculture Course*, p. 94.

out, our butcher skillfully separates the peritoneal tissue from the sides of the carcass so that the whole abdominal or peritoneal sack can be removed intact, still containing the abdominal organs. The gallbladder is removed first so the bile doesn't contaminate the meat or the rest of the organs. Once the sack is removed, we move it to a clean area so we can separate out the intestines, stomachs, mesentery, peritoneum and other organs.

The first step is to separate the intestines from the stomachs and identify the parts that you want to use. The small intestines will be attached to the Isles of Mesenteria, which holds them all together. The large intestines will be darker in color and harder to excise from the other mesenterial tissue holding them together. If you were making only a small number of preparations, the pinkish part of the small intestine would be preferred. Steiner's direction was to use the part that is used to make sausages for human consumption.

Left: Separating the small intestines from the Isles of Mesenteria
Right: Small intestines have been almost all removed;
what is left is large intestines and the Isles of
Mesenteria, the fatty tissue at bottom of photo.

We use all of the intestines, small and large, for the quantity of preparations needed at JPI. With a sharp knife or razor blade, make an incision in the mesentery to carefully separate it from the intestines. Leave some of the mesenterial tissue attached to the intestines so you do not get too close and puncture the intestines. Decide where you are going to make your cut across the intestines. Before you cut through the intestine, tie a string around both sides of where the cut will be. This is to prevent the contents of the intestines from leaking out everywhere when you make the cut. Proceed to separate a manageable length (5 to 6 feet) of intestine from the mesentery, always staying a good half-inch or so away from the intestinal wall to avoid weakening it or puncturing it. Once you have separated the desired amount, make two tie-offs again before cutting off this section.

Continue to repeat these steps until you have the amount of intestine needed. The next step is to gently clean the food contents from the intestines. Undo one of the tie-off strings and put that end of the intestines into a bucket and loosely squeeze the contents out. Remove the string from the opposite end, insert the nozzle of a hose and run a quart or two of water through the intestines, just a quick flush. You only want to remove the rough food contents to make room for the chamomile. You do not want to remove all the chyme (liquefied and digested foods, hormones, bacteria, enzymes, bile, and other digestive juices) that is on and between all the thousands of finger-like projections or folds called villi that line the interior wall of the intestines.

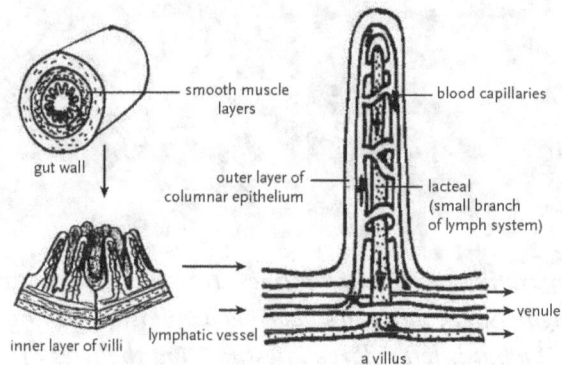

Anatomy and physiology of the cow's wall of small intestine showing villi (drawing by Ruth Lawson, Otago Polytechnic)

These villi greatly increase the interior surface area of the intestines, some sources say by up to six hundred times. The residual chyme contributes to the transformation of the chamomile and the characteristics of the final preparation. You definitely would not want to clean out the intestines as thoroughly as you would if you were making sausages for human consumption.

Some practitioners in Switzerland prefer to use the caecum (cecum), which is the beginning of the large intestines. It is a pouch with only one opening, which connects the small intestines to the ascending colon of the large intestines. The appendix is attached to it. It would be an interesting research project to analyze the various parts of the intestines for enzymes, yeasts, hormones, bacteria, and such, to see what, if any, difference there might be, and then test the resulting preparations made from these parts.

The intestines should be stuffed as soon after slaughter as possible. If you are unable to stuff them within three days, they should be frozen until used. When we have to freeze them, we put them in a large plastic food freezer bag with enough water to cover them and seal the air out. With the water surrounding them, the tissue does not break down but becomes freezer burned. If you do not freeze them, the digestive process continues and the decay process starts very quickly. They will get very stinky, putrid, and unpleasant to work with after the second day, even kept in the refrigerator underwater. The decay process also weakens the wall of the intestines making them difficult to stuff tightly. If you are obtaining the intestines from a butcher, you would want to pick them up the same day as slaughter, take them home, separate them from the mesentery, clean them gently and either use them right away or freeze them. (We freeze the deer bladders this way as well, in a plastic bag or in a yogurt container, always covering with water, until we accumulate a supply and have the time to inflate them to dry.)

Stuffing the Intestines

AP: *How do you prepare the intestines for stuffing?*

HC: First, moisten the blossoms as I previously described. Then cut the intestines into nine-inch sections and remove the fat. Keep the rest of the intestines in a bowl of water until you are ready to use them so they don't dry out. The fatty tissue must be removed from the side of the intestine. There are a couple of ways to do this but it generally works better to do this with a razor blade before they are stuffed. The nine-inch section is then spread out flat across a cutting board with the white fatty mesentery tissue stretched away from the intestine. The razor blade is then used very carefully to scrape and cut away all this fatty tissue without removing intestinal tissue or cutting through the wall. If we do not remove this fatty tissue, a little grub worm about a half-inch long will show up in the finished preparation. At the annual JPI fall workshop, which is hands-on, everyone gets gloves, a wooden cutting board, a razor, and one or more intestine sections to stuff.

 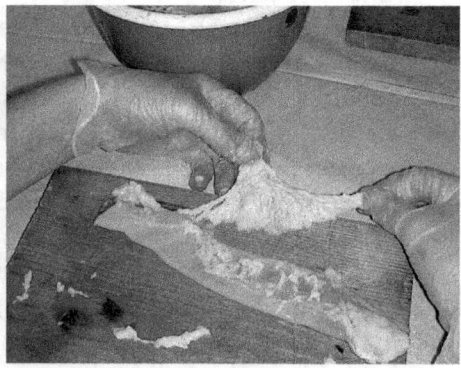

Left: Fatty tissue attached to the side of the intestine
Right: The fatty tissue has been separated from the intestine.

AP: At the workshop, I was trying to do a good job to get all of the fatty parts off. When I was stuffing, I discovered I had taken off too much tissue, which resulted in a thin, weak area. I wasn't able to stuff the rest of the sausage as tightly without breaking through. Some other people did get holes and they had to stop stuffing, ending up with a much shorter sausage.

HC: An alternative way to remove the fat is to stuff the sausage and let it dry for a couple of days. Then you can rub or peel the fat off. One still needs to remove at least an inch of the fat from both ends before stuffing because you can't get to it when the ends are closed off. Next, the one end of the intestine is tied off with a string and the other end is opened up to stuff the blossoms in. Some years ago, we found some sausage stuffing funnels with a wide opening. Unfortunately, they are no longer available, and other funnels we have tried are too narrow. Anyway, we insert the end of the funnel into the intestine and push the chamomile, which has been adequately moistened, through the funnel with a dowel. The blossoms are worked down to the bottom of the intestine tube and packed in tightly with your thumbs. When you have stuffed up to about an inch or less away from the top, one side of the open end intestine is folded over the blossoms. The other side is then folded over the first and all the edges are pressed down tightly. This will seal the end closed and you don't need to tie it shut with a string.

AP: My mother usually stuffed longer lengths of intestines. Is there a reason that you cut the intestines into nine-inch lengths?

HC: It is harder to get the chamomile into a longer tube and harder to get it packed tightly. Given Steiner's suggestion for packing the manure more tightly in horns—hammering it in, to make it more effective for the American continent—I have applied that insight to the other preparations, as well, including chamomile, dandelion, and yarrow.

"You use these airy, flying-away forces, the scent that wants to go away from the earth, and you bring it into the most earthly, destructive surrounding, the wall of the intestines.... In these 'sausages' that you have made, the calcium processes are brought together. The exhalation process of camomile is necessary so that the central and centralized forces of calcium can really be brought to the compost heap and from there to the soil. You catch the 'desire of calcium' and you surround it by what contains our earthly nutrition stream and destroys it. These you expose during winter to the cosmic summer forces under the soil." —KARL KÖNIG[3]

---

3   König, *Earth and Man*, p. 304.

Stuffed chamomile "sausages" ready to be buried

Burying the Preparation

AP: Is the chamomile buried the same way as the dandelion?

The pit is lined with wire mesh. Stuffed chamomile "sausages" are safely covered with wire mesh before being covered with soil and buried for the winter.

HC: Yes, it is buried in a pit twelve to eighteen inches deep. Our pits are about fourteen to fifteen inches deep. You want to stay in the fertile layer of the soil. We line the pit with half-inch mesh hardware cloth (galvanized steel) that is big enough to fold over from two sides to cover the sausages. We frequently use two pieces placed on top of each other, at right angles, so we can fold them over from all four sides. Next, we put in about

503: Chamomile: The Healer for People and Plants

a half-inch of peat moss to prevent the metal from touching the fiberglass screening. We place fiberglass screening over that, again using a large enough piece that can be folded over from all sides. The sausages are placed on the screening. We arrange them in concentric circles or a spiral, starting in the center. The screening is folded over to make a package, making it easier to retrieve the finished preparation without losing any precious material. Another layer of about two inches of peat moss is put in over the fiberglass screen before folding the hardware cloth over the top. The hardware cloth prevents animals from digging up the preparations and also prevents shovel damage to the screen and preparations when you dig them up in the following year. This method of protecting the sausages is advisable. You would be amazed at the number of people who call me up, desperate because the fox ate their chamomile or dandelion, or a dog dug it up, or alternatively, they can't find it. The pit needs to be clearly marked so you can find it in the spring as well.

AP: *Is there a preferred location for burying the chamomile?*

HC: Steiner does say to choose a location where the snow will remain for a long time and where the sun will shine on the snow. Given this part of Virginia, we often don't have very much snow. When we do get a decent amount of snow, somebody goes down to the pit and shovels some more snow on top. This is to make sure there is plenty of winter snow-water melt seeping down into the pit. If you have winter rain, you will get winter moisture. The snow, being exposed to the sun for several days, will collect sun and other cosmic forces. This is not what happens when it rains.

AP: *So the people in southern locations or in the tropics are out of luck?*

HC: I'm hopeful that the winter rains would help compensate for that. The sun is stronger closer to the equator so maybe that would suffice, especially if it rained during the daytime. Steiner gave the agriculture lectures to farmers in Europe where there was snow. He didn't live long enough to give advice for other locations. His eight lectures were hints and beginning indications. He encouraged farmers to experiment, observe, and research.

AP: When are the chamomile sausages buried and how long do they stay in the ground?

HC: Steiner only says for them to be in the earth throughout the winter when the earth is most inwardly alive. I prefer to get them in the ground before Advent (at least four Sundays before Christmas). I have observed the resulting preparation to be of better quality if it is buried before Advent. If you are farther north, you may want to get them in the ground much earlier, before it snows and before the ground freezes. We usually keep them in the ground for about six months. Sometimes, it comes down to getting to it when you can get to it. I prefer to wait at least until Ascension Day (thirty-nine days after Easter Sunday) before digging up any of the preparations. We have chamomile buried in two pits, side by side. This year we dug up one in late spring. The other one will be dug up at the fall workshop so participants can experience the unearthing as well as the burial. They can observe, smell, and feel the chamomile before it goes into the earth as well as when it comes out of the earth, transformed by its time in the intestine, exposed to the cosmic winter forces.

AP: Are the intestines still evident when they are dug up?

HC: There is generally some intestinal tissue still in evidence, but usually you cannot pick up a whole sausage intact.

AP: What are the next steps, after they are removed from the earth?

> "Add them to the manure.... You will thus get a manure with a more stable nitrogen content, and with the added virtue of kindling the life in the earth, so that the earth itself will have a wonderfully stimulating effect on the plant growth. Above all you will create more healthy plants—really healthier."
> —Rudolf STEINER [4]

---

4 Steiner, Agriculture, p. 94

HC: We gently lift the screen, containing the chamomile preparation, out of the ground and put it into a tray to transport it to the work area. There, we crumble up the sausages and gently rub them through a quarter-inch mesh screen to attain a consistent texture. The screening process also separates out the strings that were used to tie off the intestines and any pieces of intestinal tissue that have not decomposed. In the finished preparation, the blossoms will be completely transformed into a dark humus-like material, no longer identifiable as chamomile. If the blossoms are still evident and not broken down, several pieces of the intestinal tissue that were just screened out can be placed in the crock with the preparation. The intestinal forces will continue to radiate into the material and continue the breakdown process. Once or twice, several years ago we had to do this, as the chamomile was not broken down as completely as it should have been.

Storage

AP: How are the preparations stored?

HC: The preferred storage container is a glazed ceramic crock with a loose-fitting lid of ceramic, slate, or glass. A glass container may also be used. The crock is placed in a lidded, wooden box, and is surrounded on all sides, including the bottom, by at least 3 inches of granular peat moss. A pillowcase or feed sack can be filled with peat moss to create a pillow to insulate the top of the crock. The peat insulation serves to keep the radiating forces of the preparations contained in the crock so they don't dissipate and also to help prevent adverse radiations from penetrating into the preparations. I have to recommend that you do not put all of your preparations in the same box. I have found that if you do, somehow, the other preparations seem to affect both the chamomile and the dandelion. They will dissipate, just disappear. The chamomile and the dandelion preparation each need to have their own separate box. I keep all of the preparations in their own boxes now. The preparations should be checked at least monthly to ensure that the moisture content is stable and that no critters are getting into the preparations or mice are nesting in the peat moss.

The finished preparation is unearthed.

Effects of the Preparation

AP: *What is the effect of BD503 on the compost pile, the earth, and plants?*

HC: The chamomile preparation is related to the earthly nutrition stream and strengthens the life body of an organism, be that organism a plant, the compost pile, or the earth. With its association with limestone or calcium, it helps stabilize the nitrogen in the compost pile and helps the soil and plant absorb more nitrogen from the atmosphere. Both chamomile and the intestines are connected to Mercury, which is always in motion, creating chaos and lacking form. This is the polarity of Jupiter, the silica process, and BD506 (dandelion) that creates order, form, and structure. The outer planets bring in the cosmic nutrition stream, and the inner planets enhance the earthly or more physical nutrition stream. The relationship of Mercury to chamomile is best explained by C.B.J. Lievegoed in his pamphlet, "The Working of the Planets and the Life Processes in Man and Earth."

AP: *Have you experimented with using the chamomile preparation by itself on plants?*

HC: I can't say that I have. Other than stuffing sausages, which is a lot of fun really, I haven't been as drawn to chamomile as I have to some of the other preparations—oak bark, in particular, and yarrow. Recently, someone called who has a problem with cocklebur weevil infesting his dahlias. The weevil lays eggs and the larvae burrow into the stem of the plant. It came to me that the situation calls for the earthly nutrition stream to be strengthened. So I suggested that he stir and spray the chamomile preparation along with milk and honey. He had used all of the preparations already, so I was looking for something specific. We'll see what happens.

Steiner speaks of the harmful effects of fructification being countered by chamomile and its relationship to calcium. I have taken this to mean that if the ripening processes were too excessive or occurring too early they would be harmful. This would be the case with corn smut, for instance. If earthly forces are not strong enough and the cosmic are too strong going up the plant, you have to bring in a counterforce with the earthly influence of the chamomile preparation.

> "Chamomile, however, assimilates calcium in addition (to potash). Therewith, it assimilates what can chiefly help to exclude from the plant those harmful effects of fructification, thus keeping the plant in a healthy condition." —Rudolf STEINER [5]

AP: *Would the early bolting of leafy plants be an example of too much cosmic influence, bolting before there is very much leaf growth?*

HC: That can happen. Ideally, everything needs to be in balance and harmony so that these forces work in the plant the right way. To counter bolting, one would build up the earthly forces through the application of BD500 for sure, and possibly yarrow and certainly chamomile. They would all strengthen the earthly. You have to find avenues to stimulate both the cosmic and the earthly nutritional streams.

The cosmic nutrition stream, which is basically silica and the forces from the outer planets, comes into the earth and impacts

---

5 Ibid.

silica rock deep in the earth. These forces radiate or flow back up through the earth into the plant to carry the silica forces into seed formation. The opposite of this is the chamomile, which strengthens the compost so that the earth will be more alive and the life force of the earth can also come up into the plant. By stimulating both these forces with the preparations we help the earth to be more alive, which results in healthier plants.

> "Mercury: Opposite to Jupiter stand the Mercury forces. Where Jupiter gives rise to harmony and order, Mercury creates chaos, though not an ordinary one, but one, which we might call 'sensitive chaos'—movement without direction, but ready to flow into anything that might be suggested from the outside. Mercury is streaming movement, adapting itself to any resistance it may find.... The one thing Mercury never abandons is movement, in this streaming and flowing. Mercury is active in the Camomile Preparation 503, which stimulates plant growth through potassium and calcium. The treatment with the intestines intensifies the Mercury activity.... Even today Camomile is widely used in medicine, mainly in the whole region of mucous membranes. Wherever there is inflammation, i.e. where chaos prevails, where mucus and pus are produced, Camomile soothes, deodorizes, cleanses. The main sphere of application, however, is the intestines. We see that Camomile in the first place has a strong effect upon the intestines, secondly, it brings what is congested...into motion." —C. B. J. LIEVEGOED [6]

AP: Is there anything you would like to add?

HC: The cosmic and earthly nutrition streams are key to solving the riddle of the Agriculture Course. Steiner repeatedly talks about polarity. Understanding what each of the preparations is doing, in relation to either the cosmic or earthly nutrition stream, is crucial to understanding the whole Agriculture Course, I think. It's a big mystery, and I'm still wrestling with it.

AP: Thank you, Hugh.

---

6   Lievegoed, "The Working of the Planets and the Life Processes in Man and Earth."

504

Urtica Divica
Stinging Nettle

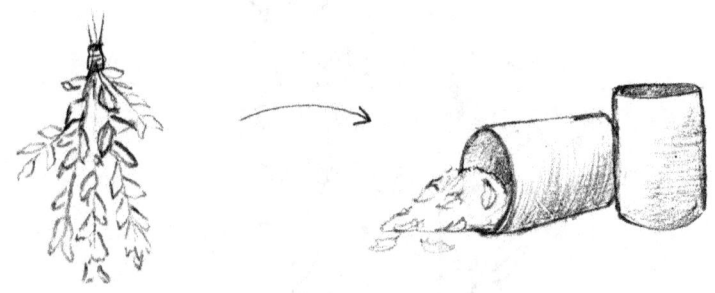

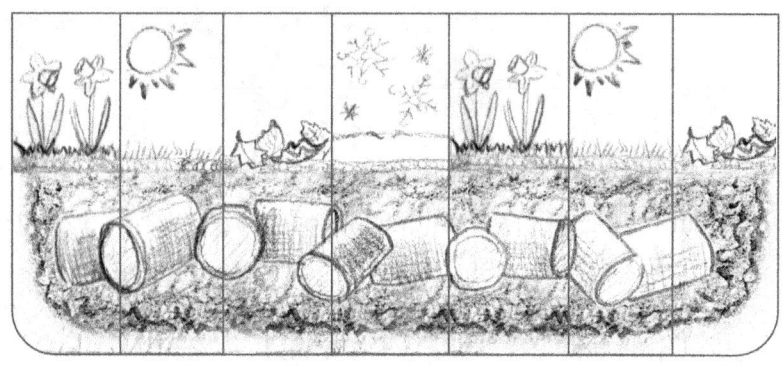

504: The Stinging Nettle Preparation

Hugh J. Courtney
Applied Biodynamics, *no. 24, fall 1998*

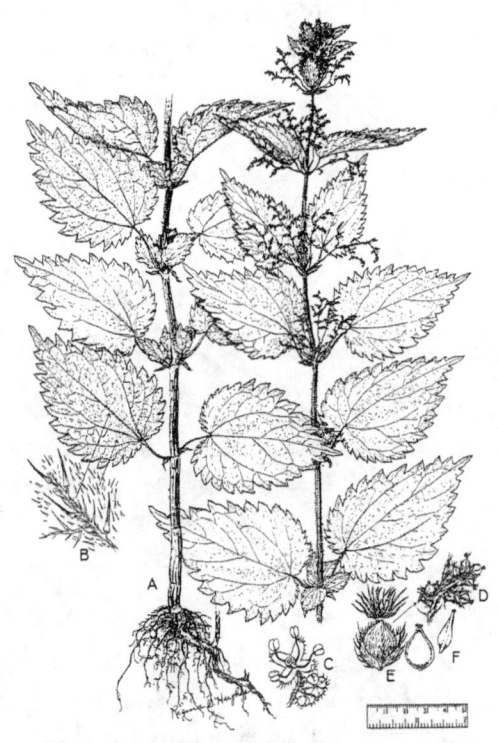

To begin, we examine the unique plant from which this preparation is derived. Steiner confirms the uniqueness of stinging nettle by referring to the fact that while substitutes may be found for the other preparation plants, specifically yarrow and chamomile, there is one plant whose beneficial influence on the manure is such that it would be next to impossible to find a substitute for it.

504: The Stinging Nettle Preparation

We are often not very fond of this plant, at least not in the sense of wanting to fondle it, because the plant in question is stinging nettle. When Steiner is subsequently questioned as to whether it is the annual or the perennial stinging nettle that is to be used, he responds simply with the technical name of the plant, *Urtica dioica*, that is, the perennial stinging nettle. *Urtica* is from the Latin *urere*, "to burn," which one presumes is in deference to the stinging hairs of the plant. *Dioica* is from the term *dioecious* (from the Latin for "two houses"), which is a botanical designation for plants having the male and female reproductive organs borne on separate individuals of the same species. Do not be surprised, however, if you sometimes observe a single plant that has both a distinct zone of male flowers as well as a distinct zone of female flowers. In other words, it is monoecious (both sexes in "one house"). Jim Duke, a renowned ethnobotanist, once pointed out in a private conversation, that sometimes "the plants don't bother to read the book, and do whatever is necessary to continue the species."

The inflorescences are found in the axils of the upper leaves. The male inflorescences tend to project out sideways, while those of the female are more pendulous. A U.S. Department of Agriculture publication gives the following description of *Urtica dioica*.

> Perennial *reproducing* by seeds and creeping rootstocks; stems erect, 1-2 m. tall, ridged, bristly-hairy with stinging bristles (0.75-2 mm. long); *Leaves* opposite, simple, egg-shaped to heart-shaped, coarsely serrate, hairy or glabrous, with or without stinging bristles, usually twice as wide as the length of the petiole, the stipules linear-lanceolate, 5-15 mm. long, green to pale brown, minutely pubescent; *Inflorescence* branched, many-flowered, loose to dense, panicled spikes; *Flowers* mostly dioecious, small, greenish; Staminate flowers with 4 perianth segments and 4 stamens; *Pistillate flowers* with 4 perianth parts and a 1-celled ovary; *Fruit* (achene) 1-1.5 mm. long, flattened, egg-shaped, minutely glandular, yellow to grayish-tan, the calyx and remnant of the style often persistent. June-September.
>
> Waste places, roadsides, vacant lots, rich soil, and edge of damp woods. Naturalized from Eurasia. Throughout all the

United States excepting southern Georgia, most of Florida, and that part of the United States from northwestern Washington through most of Texas.[1]

It is important to compare what Steiner says about the constituents of stinging nettle, to try to determine what qualities it possesses that make it so unique among plants. He goes so far as to define that uniqueness in the following statement: "But stinging nettle is in fact the greatest benefactor of plant growth, and it can hardly be replaced by any other plant. If it is not available locally, you really must get the dried herb from somewhere else."[2] It should be kept in mind, however, that when Steiner speaks of an element such as sulfur, or calcium, he is not so much referring to the gross element or chemical substance as he is to the process associated with that substance; or, even more exactly, the spiritual being behind that substance. Since most of us are not as gifted or adept as Steiner was, and cannot hold a conversation with such beings, we must exercise our imaginations mightily in order to come to grips with what he is saying as he speaks of these qualities or constituents of stinging nettle. In Steiner's description, he says:

> Stinging nettle is a real jack-of-all-trades; it can do many different things. It too contains sulfur, which, as I have already explained, plays an important role in assimilating and incorporating the spiritual. Stinging nettle also carries the radiations and currents of potash and calcium, but in addition, it has a kind of iron radiation that is nearly as beneficial for the whole course of nature as the iron radiations in our blood are for us.[3]

When we resort to a modern scientific view of stinging nettle, we are given a clinical description that reads as follows:

---

[1] USDA, *Common Weeds of the United States.*
[2] Steiner, *Agriculture*, p. 98.
[3] Ibid.

504: The Stinging Nettle Preparation

Dried, young pre-flowering plants contain, per 100 g. 30.4 g crude protein, 3.4 g fat, 10.3 g cellulose, 39.6 g N-free extract, 16.3 g ash, 2,970 mg Ca, 680 mg P, 32.2 mg Fe, 650 mg Mg, 3,450 mg K, 140 mg Na, 4.3 mg Mn, 540 mg S, 680 mg Si, 270 mg Cl, and 20.2 mg beta carotene. Oil from the seeds contained 11.5% oleic-, 73.7% linoleic, 1.7% linolenic-acid, and *circa* 7.0% saturated acid (mainly palmitic), 4.5% glycerol, and 1.6% unsaponifiable material. Fresh plant material contained 80 µg vitamin B1 per 100 g and 15.7 mg chlorophyll. Betaine, choline, and lecithin occur in the leaves. Carbonic, formic, and silicic acids are also reported, with phytosterins and tannin.[4]

While the chemical analysis readily supports Steiner's characterization that the plant contains potash and calcium "radiation and currents," the limited amount of iron and its "sister" element manganese makes it difficult, from a substance point of view, to conclude the same degree of importance that Steiner gives to the iron radiation. It is also interesting to note that stinging nettle actually contains more of the other "light bearing" element, phosphorus, than it does of the sulfur that Steiner assigns such an important role in "assimilating and incorporating the spiritual." Obviously, the chemical analysis, although useful, fails to provide us the necessary understanding of this plant that Steiner possessed. If we observe the soil around the roots of the stinging nettle, even if it has been in place for only a short time, we cannot help but marvel at how dark, even black, the soil is. Although I have never seen an analysis of the humic acid or humus content of such soils, it is easy to believe that they would be much higher than the soil under most other plants nearly in the same basic parent soil. While I hesitate to draw any particular conclusions in comparing Steiner's spiritual-scientific analysis with a chemical substance analysis at this point in my efforts to understand it, I believe it is worthwhile to provide such a comparison so that others besides myself may reflect upon it as well.

---

4   Duke, *CRC Handbook of Medicinal Herbs*, p. 502.

When using the plant itself in our gardening efforts, a freshly made tea, diluted one in ten, and sprayed on other plants can be effective in combating aphids. On the other hand, if we make a tea of stinging nettle and allow it to ferment for 48 hours or more, and use it as a spray on plants, again using a 1:10 dilution, it serves as a strong, nitrogenous type of fertilizing agent. The sulfurous character of the odor that emanates from this fermented nettle brings us closer to a realization of the sulfur "process" that Steiner may have been referring to in the quote above. When using this nettle tea ferment, one actually has to be somewhat cautious, as overuse, say, more than twice in a growing season can cause the plants sprayed with it to be more subject to fungus attacks, in the same way that happens when one uses too much raw, uncomposted manure on plants.

When we approach the making of the preparation itself, it has been my observation that there has been a certain carelessness that has crept into the business of making this particular preparation. To properly delineate that carelessness, let us return to the source and quote directly from Steiner's indications for making this preparation:

> Now, to improve your manure still more, take whatever stinging nettles you can gather and once again let them wilt slightly. Then compress them a bit and put them straight into the ground without any bladder or intestines, though you can use a thin layer of loose peat or something similar to separate them from the surrounding soil. Bury them right in the ground, but mark the place carefully so you do not just dig out plain soil when you come back for them. Let the nettles spend the winter and also the following summer in the ground; they need to be buried for a whole year. Then you will have a substantiality that is extremely effective.[5]

The general practice amongst a very great number of biodynamic preparation makers is to harvest the stinging nettle in June or July, bury it, and then dig it up a year later in the following June or July

---

5   Steiner, *Agriculture*, p. 99.

504: The Stinging Nettle Preparation

(in some cases, it is dug up even earlier in April or May). In other words, most preparation makers are abiding by Steiner's indication that this preparation should be buried for a whole year, but have not paid sufficient attention to his statement that it is to "spend the winter *and the following summer in the ground.*" I have long since come to the conclusion that Steiner chose his words very carefully in presenting to the world the wealth of spiritual-scientific insights that he gave, and that if he meant this preparation to also spend the "following summer" in the ground, it is for a sound and important reason. I will reserve identification of that for a subsequent section of this article after I have described some of the actual practices and techniques that have evolved over the years in my making of this particular preparation.

In practice, most people attempting to make this preparation, especially in soils that have been treated biodynamically for a while, and that therefore have a very high earthworm population, have long since learned that to surround the wilted nettle with "a thin layer of loose peat," is simply not sufficient to ensure that you will recover any of the finished preparation a year or longer after you first have buried it. In more northerly climates, such as in Canada, some have found it sufficient to bury the stinging nettle in a burlap bag surrounded by peat moss. In more active southern soils, the earthworm activity is so great that one would be very lucky to recover any evidence of the stinging nettle preparation, or even a scrap of nettle permeated burlap. The practitioner should understand that earthworms absolutely love to feast on this preparation. As a consequence of such experiences of not being able to recover any of this preparation, a number of techniques have evolved to assure that at least some of it can be recovered. The most commonly used technique is to make a "pillowcase" out of window screening into which one packs the harvested and wilted stinging nettle. The accompanying diagram illustrates this technique. What one needs for this procedure is window screening, either fiberglass or ferrous-based screening (never copper or aluminum), and a good heavy-duty stapler to staple or stitch up the pillowcase. Three sides of the pillowcase are stapled together using a double fold and a generous number of staples. The nettle is placed inside the pillowcase screening and the fourth side is stapled closed. In a pit twelve to eighteen inches deep, the nettle pillow is then placed on a bed of milled peat moss at least three inches thick and the pillow is covered with another three-inch-thick layer of peat on top and around all the sides. The excavated soil is used to cover the contents of the pit so that at least twelve inches of soil is covering the preparation.

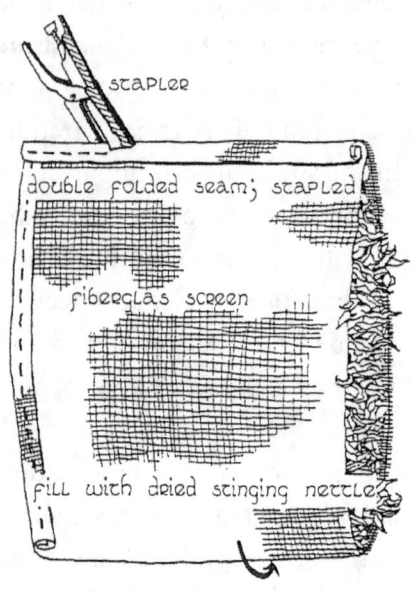

When this technique is used one is likely to recover a fairly high percentage of good, finished BD504, although even here, diligent plant roots can find their way through the screening and thus provide a pathway for even more diligent earthworms. If one harvests and buries the stinging nettle in June or July, it should not be dug up until a year from the end of the coming month of September. If a second crop is harvested and buried in September, it would be dug up approximately a year later at the end of summer in the following year.

Another technique, which we have been employing to some extent in recent years, was prompted by a conversation with Alex Podolinsky when he paid a visit to Kimberton, Pennsylvania, over a period of several days in June 1992. His technique for ensuring recovery of this preparation involves packing it into a length of clay pipe such as was formerly used in drain fields for septic systems and agricultural drainage. Once we located a few drain tiles, we began to employ this

504: The Stinging Nettle Preparation

technique and thereby achieved quite a vibrant nettle preparation in the end (though I must confess to a certain skepticism when I first tried this method). The materials we use for making the nettle preparation according to this technique are as follows: one or more clay drain tiles, a piece of wood approximately three-quarters inch in thickness cut to fit the inside diameter of the tile, window screening also cut into circles to fit inside the tile at the rate of two pieces per tile, a supply of long-fibered sphagnum peat moss,[6] a supply of moist clay to seal the ends of each tile, a wooden tamper, a two-pound sledgehammer, and finally, of course, a supply of harvested stinging nettle, either fresh or dried.

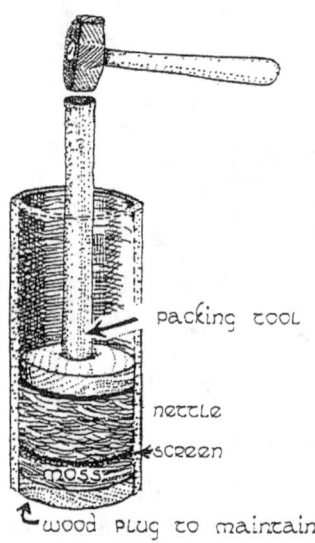

One begins by placing the circular piece of wood at the bottom of the upright drain tile, followed by a good handful or two of the long-fibered sphagnum moss that has been moistened and wrung out. Using the wooden tamper and sledgehammer, tamp the peat moss firmly to form a one or two-inch thick plug at the bottom of the tile, and then place a circle of window screening on top of it. The next step is to take a handful of chopped stinging nettle, place it on top of the screening, and pound it down tightly with the hammer and wooden tamper. Continue this process until the upright drain tile is fully packed

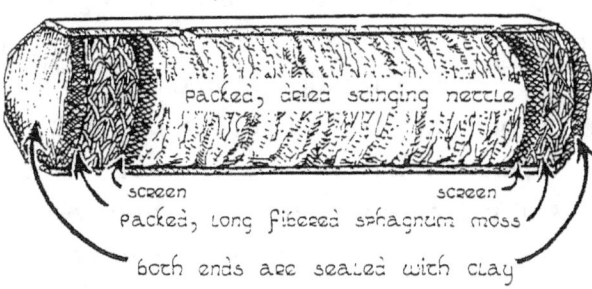

---

6   Long-fibered instead of milled peat moss is used since it will, when moist, adhere into a plug when tamped down.

with stinging nettle to within three inches or so of the top. At this point, place the second circle of screening on top of the nettle. Take another handful of long-fibered peat moss and proceed to make another one or two-inch thick plug, leaving space about half an inch from the top. Finally, remove the circular piece of wood from the bottom of the tile, and fill the cavity at both ends with moist clay to seal the tile. When placing the stinging nettle in the tile, one may choose to moisten it or not, but the end product seems to be somewhat better if it is not moistened (even if using the dried herb to make the preparation).

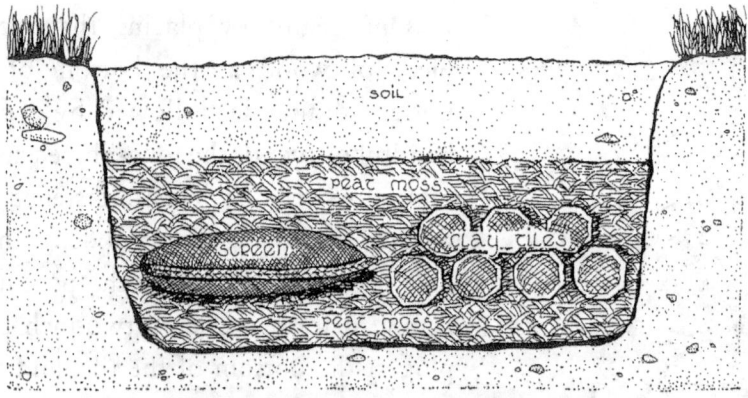

The statement ending the last paragraph may well bring up the question of how one judges the quality of the finished preparation. While it is particularly difficult to quantify and measure the quality of any of the preparations in a totally objective way, there are certain signs that provide clues to the quality of BD504 when one digs it up. Initially, however, when digging up the preparation and examining it, it is a good practice to suspend all judgment and simply observe the preparation very closely, essentially without allowing any thoughts of judgment to enter into the observation. I call this technique "seeing into the preparation." After meditating/observing the preparation in this fashion, I will shortly have an awareness as to the quality of this particular preparation with

504: The Stinging Nettle Preparation

respect to previous years' results. Over the years, however, I am always pleased to see an end product, black in color as can be, for my perception is that the blacker the end product is, the higher the quality it is likely to have. Any signs of life in the form of almost microscopic white bugs (which, to date, I have yet to identify), is also a signal to me of positive quality. Additionally, if one takes a small pinch of the finished preparation between one's thumb and finger and rubs vigorously, one will get a greater or lesser degree of smearing. The more the pinch of stinging nettle disappears or disintegrates into a stain or smear between thumb and finger, the higher the quality it has in all probability. Ultimately, of course, the answer as to the quality of any given preparation can only be found by using it and seeing the results obtained. With respect to the stinging nettle preparation, one might consider using a bit of it as a seed bath for a dozen or so grains of barley (barley is the only seed for which BD504 is suggested as a seed bath), using as a control another dozen grains in a seed bath of plain water and comparing the results to see if the preparation made a difference in germination time. Beyond this, one should simply make compost and see if the results are pleasing and if the plants grown with the compost are prospering.

The stinging nettle preparation can be viewed as having a hot, dry character. Culpepper and other noted herbalists assign these same characteristics to the plant because of its relationship to the planet Mars. Separately, BD504, when applied as a spray or as a watering-in agent around plants, appears to have an ability to strongly stimulate leaf growth, as well as to be of some benefit in preventing frost damage. Much work lies ahead in the future to determine what effects using this preparation might have on plants in biodynamic culture when used by itself.

The stinging nettle preparation, truly deserves to be known as the Michaelmas preparation. This view arises from Steiner's indication that this preparation should spend the winter in the earth and remain there through the following summer. Other than BD501, Steiner did not give instructions for any of the other preparations to be buried during any part of the summer.

Having long since learned that Steiner did not indulge in careless word choices, I feel that in such cases like this he should be taken with total seriousness. In other words, if he said it should be buried through the following summer, he meant it. I have been unable to determine how the practice arose of digging up the stinging nettle preparation at the beginning of summer or even as early as April or May after its wintertime burial. I can only conclude that this widespread practice has diminished the full effectiveness of this preparation, and is at least one of the factors involved in the relatively insignificant impact that biodynamic agriculture has had throughout most of the world.

When one begins to look for the reasons behind this statement of Steiner about the stinging nettle preparation, the lectures he gave on the Archangel Michael (pronounced "Mick-ah-el") provide us with the needed insight. Although my efforts to interpret these lectures as they might apply to the stinging nettle preparation may fall short of the mark, please bear with me, as it takes us rather deeply into the body of Steiner's esoteric work. Steiner speaks of a certain inner process that takes place within the human being between St. John's Tide (June 26) and Michaelmas (September 29), which also has a parallel in events taking place in the outer physical world, and in spiritual dimensions as well. Within the human being, there takes place an inner process of combustion, but we must not picture this as being like external combustion. All the processes that take a definite form in the outer world also occur within the human organism but in a different guise.[7] Steiner continues:

> The inner process that occurs during high summer is a permeation of the organism by what is represented crudely in

---
7   Steiner, *The Archangel Michael*, p. 219.

504: The Stinging Nettle Preparation

the material world as *sulfur*. When we live with the summer sun and its effects, we experience a sulfurizing process in our physical-etheric being. The sulfur that we carry within us as a useful substance has a special importance for us in high summer, an importance quite different from its importance at other seasons. It becomes a kind of combustion process. It is natural for humans that the sulfur within us should thus rise at midsummer to a specially enhanced condition. Material substances in different beings have secrets not dreamt of by materialistic science.[8]

Although this is a process seen only with a spiritual vision on the order of Rudolf Steiner's, it has at least two major spiritual consequences for both human beings and for events within the cosmos. Steiner tells us that this inner sulfur process "has a great importance to the evolution of the cosmos when in summer human beings shine inwardly with the sulfur process."[9]

Besides the glowworms or lightning bugs that shine out around St. John's Day in the physical realm, "the inner being of humans then begins to shine, becoming visible as a being of light to the etheric eyes of other planetary beings. That is the sulfurizing process. At the height of summer, human beings begin to shine out into cosmic space as brightly for other planetary beings as glowworms shine with their own light in the meadows at St. John's Tide." While this sulfurizing process within the human being is a positive, perhaps even nourishing process as far as these other spiritual and/or planetary beings are concerned, it does have a negative aspect or consequence for human beings. Steiner defines this negative consequence thusly: "…but at the same time, it gives occasion for the Ahrimanic power to draw near to humanity. For this Ahrimanic power is closely related to the sulfurizing process in the human organism."[10]

---

8   Ibid., pp. 219–220.
9   Ibid., p. 220.
10   Ibid.

Here reference is made to the spiritual being Ahriman, who has been characterized by Friedrich Rittelmeyer as the "spirit of materialistic intellectualism."[11] Elsewhere, Steiner tells us to "think of everything that presses us down upon the earth, that makes us dull and philistine, leading us to develop materialistic attitudes, penetrating us with a dry intellect, and so on; there you have a picture of the Ahrimanic powers."[12] These Ahrimanic powers, by virtue of the sulfurizing process within human beings, attempts "to ensnare and embrace them, to draw them down into the realm of half-conscious sleep and dreams. There, caught in this web of illusion, human beings would become world-dreamers, and in this condition, they would be prey to the Ahrimanic powers. All this is significant for the cosmos also."[13] Since most human beings have not reached the necessary level of self-consciousness to enter into the deeper spiritual meaning within outer nature of the processes of dying and fading that are associated with the Michaelmas season, we have a clue as to the strength of the materialistic intellectualism that pervades humanity so thoroughly today. We do, however, have some help available both in inner human processes, as well as outwardly in nature. Steiner tells us:

> And when in high summer, from a particular constellation, meteorites fall in great showers of cosmic iron, then this cosmic iron, which carries an enormously powerful healing force, is the weapon that spiritual beings bring to bear against Ahriman, as he seeks Dragon-like to coil round the shining human forms. The force that falls on the earth in the meteoric iron is truly a cosmic force...to gain a victory over the Ahrimanic powers, when autumn approaches. And this majestic display in cosmic space, when the August meteor showers stream down into the human being shining in the astral light, has its counterpart—so gentle and apparently so small—in a change that occurs in the

---

11 Rittelmeyer, *Rudolf Steiner Enters My Life*, p. 124. Friedrich Rittelmeyer was the first leader of The Christian Community, an international movement for religious renewal founded with the cooperation of Rudolf Steiner.

12 Steiner, *The Archangel Michael*, p. 103.

13 Ibid., pp. 220–221.

504: The Stinging Nettle Preparation

human blood. This human blood is permeated throughout by impulses from soul and spirit, is rayed through by the force that is carried as iron into the blood and wages war there on anxiety, fear, and hate. The processes that are set going in every blood corpuscle when the force of iron shoots into it are the same on a minute human scale, as those that take place when meteors fall in a shining stream through the air. This permeation of human blood by the anxiety dispelling force of iron is a meteoric activity. The effect of the raying in of the iron is to drive fear and anxiety out of the blood.[14]

So, we have a connection between an iron process within the human being that counters an inner sulfur process, and, in the outer universe, the August meteor showers known as the Perseids (originating from the constellation Taurus/Bull) that counter the outer sulfur process in nature. Steiner calls upon us to make conscious use of the meteoric-force in our blood. We must learn to keep the Michael Festival by making it a festival for the conquest of anxiety and fear, a festival of inner strength and initiative—*a festival for the commemoration of selfless self-consciousness.*[15]

When one ponders the picture Steiner gives us of this event in the heavens and its counterpart within the human being, along with the indications he gave for making the stinging nettle preparation and its relationship to iron, it makes a great deal of sense that the preparation remain buried through the summer months (after spending the winter in the earth). The summer burial allows it to absorb these meteoric iron forces showering the earth during this time of the year. It is important as well, perhaps, that one allows the living stinging nettle plant to gather these iron forces during the summer months. If one gathers nettle in the spring, makes the preparation and buries it, then digs it up the following June, one will end up with an immature preparation, even though it has spent a full year within the earth. The forces absorbed in the wintertime, when the earth is most inwardly

---

14 Ibid., p. 221.
15 Ibid., p. 222.

alive, must first be in place so that the preparation will be able to absorb within itself the meteoric iron forces available during the following summer. Otherwise, it is akin to taking a young child to a sermon or a serious lecture and expecting the child to be able to fully understand and expound upon the subject matter. Maturity must be reached before comprehension is possible.

In the same fashion, the stinging nettle must be exposed to the certain forces of winter within the earth, before it can fully comprehend the meteoric iron force coming to the earth during the summer months. If you prefer to make your stinging nettle preparation in June, please exercise the forbearance to wait until Michaelmas the following year to dig it up. In my opinion, however, a better course of action would be to use a June harvest of stinging nettle for making your fresh and fermented teas for agricultural purposes, or for dried nettle for livestock or human consumption. Then, at Michaelmas, use the regrowth from a second harvest of your stinging nettle to make the stinging nettle preparation. This stinging nettle will have grown back from the earlier harvest and will have received the meteoric iron force to strengthen its inherent iron radiation of which Steiner speaks. Thus, this preparation is truly anchored at both ends of its production within the Michaelmas forces. There is probably much more to be said about the stinging nettle as the Michaelmas preparation, especially when we remember that Steiner identifies the Archangel Michael as the keeper of the Cosmic Intelligence,[16] and also describes the stinging nettle preparation as the preparation that brings "an infusion of intelligence" to the soil.[17] However, exactly what Steiner meant by relating the stinging nettle preparation to an infusion of intelligence for the soil remains to be elaborated upon at another time in the future.

---

16   Ibid., p. 272.
17   Steiner, *Agriculture*, p. 100.

505

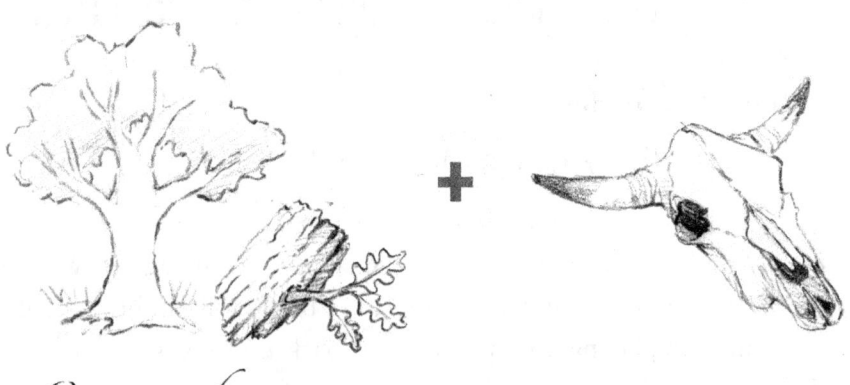

Quercus robur
Oak Bark

505: How to Make the Oak Bark Preparation

Hugh J. Courtney
Applied Biodynamics, *no. 42, 2005*

What would seem to be one of the most straightforward to make of all the preparations, the BD505, is simply described by Steiner as placing chopped up oak bark in the skull of a domestic animal. The failure to give proper and thorough attention to the making of this preparation is perhaps the greatest reason why our biodynamic efforts have thus far had so little impact on the world's agriculture. Before speaking of the challenges in the making of this preparation, let me first describe in as exacting detail as possible how we make the oak bark preparation here at the Josephine Porter Institute for *Applied Bio-Dynamics*.

Harvesting and Preparing the Oak Bark

Starting first with the business of obtaining the oak bark, the tree of choice in our area is the white oak tree or *Quercus alba*. This is the tree in North America suggested by Ehrenfried Pfeiffer as most like the English oak or *Quercus robur*, which is the particular oak specified by Rudolf Steiner when he gave the agriculture course. Apparently, Pfeiffer's recommendation was based on a more or less direct comparison of the characteristics of *Quercus robur* to the North American oak tree, *Quercus alba*, that shares a significant number of the same characteristics. Such a close resemblance between these two oaks is to be seen in the following quote from *Practical Guide to Home Landscaping*, where the white oak is described:

505: How to Make the Oak Bark Preparation

A wide-spreading, slow-growing tree with handsome branch structure, a rounded crown and leaves with rounded lobes. It grows to a height of 60 to 90 feet. Other good oaks that share the qualities of deep roots, rounded crown, slow growth and longevity are: English oak (*Q. robur*). Like the white oak, but has smaller leaves. It grows slowly and after a few generations becomes a tree of noble size and proportion.[1]

The similar species—English oak (Quercus robur), left, and the white oak (Quercus alba), right—are used to make the oak bark preparation (BD505). White oak is more commonly found in the United States while English oak is prevalent in Europe. (illustration © Mattias Baker)

No other oaks share quite the same characteristics. In an area where English oak or white oak cannot be found, Steiner's earlier indication[2] allows substitutes for all the preparations, with an exception for stinging nettle because of the difficulty of finding a substitute, giving us some leeway in choosing other members of the oak family.

In our harvest of oak bark, we remove the very outermost layer of bark, never from the north side of the tree if moss is present. We

---

1   *Reader's Digest Practical Guide to Home Landscaping.*
2   Steiner, *Agriculture*, p. 93.

look for a live, mature, healthy tree, and remove the bark from the main trunk or the largest branches if accessible. The tool of choice is a beekeepers' hive tool or a short-handled hoe, which can be pulled down against the bark to loosen it. Never do we want to penetrate as deeply as the cambium layer. Usually, a cardboard box that has had the bottom carefully sealed with tape is held just underneath the area that one is scraping with the hive tool. Alternatively, especially when harvesting a large tree, a sheet or cloth can be spread below the area where one is scraping the bark, with the bark collecting on the cloth. For the most part, a cardboard box is preferred, since it can be held closely beneath the site where one is using the hive tool, and the side of the box can be bent or shaped somewhat to conform to the curvature of the tree. When sufficient bark has been collected, we then grind it in an old corn meal grinder for small quantities or a grain grinder when dealing with a large quantity. My mentor Josephine Porter used an old hand-operated meat grinder when she taught me to make this preparation. Such equipment can often be found at farm auctions when kitchen items are included.

Left: Using a hive tool, the outermost layer of bark is harvested from a white oak (Quercus alba). Right: A small cornmeal grinder can be used to process small quantities of oak bark. (photos © Patricia Smith)

505: How to Make the Oak Bark Preparation

For greater amounts of harvested oak bark, a large-capacity grain grinder can complete the job. (photo © Patricia Smith)

IF YOU DON'T HAVE A GRINDER...

An alternative method for crushing or grinding the oak bark is suggested by Ruth Zinniker. After the bark is gathered, it is then soaked in water for a time sufficient enough to allow it to be broken up easily. After soaking, it is placed on a hard surface (concrete slab or sidewalk) and crushed with a hammer until it is well crumbled. The resultant product is rubbed through a one-quarter inch screen or sieve and the particles that will not pass through are subjected to further pounding until there is sufficient crushed oak bark to fill the skulls one intends to make.

After the grinding process, we are seeking a consistency of ground oak bark that is approximately the fineness of builders' sand used for mixing mortar. To further prepare the oak bark for placement in the skull, we brew a tea in a large pot with several larger chunks of oak bark that have not been ground. A significant quantity of oak bark tea will be needed since the ground oak bark seems to have an enormous capacity to absorb water. We want to make a good strong tea that may need to simmer for an hour or more to extract the soluble substances from the oak bark. Before actually moistening the oak bark, we also want to make sure that the tea has cooled enough so that we do not scald the life force within the ground oak bark. By ensuring that the tea is pleasantly warm to the touch of a human hand, we will arrive at the optimum temperature that is needed to moisten the oak bark.

Moisten the ground oak bark with the oak bark tea. Once thoroughly blended, the consistency is just right when a handful of the mixture can be squeezed without a drop of liquid being released. (photo © Patricia Smith)

505: How to Make the Oak Bark Preparation

The tea is added carefully to the ground oak bark and one is always surprised at the absorbing capacity of the bark. The optimum moisture content is achieved when a handful of the moistened material can be picked up and squeezed tightly in the hand with virtually no water droplets being exuded, yet the ground oak bark itself is completely moist with no dry pockets. At this point, the moistened oak bark is ready for insertion into the brain cavity of the skull of a domestic animal. So, let us now turn to the question of the skull to be used in making this preparation.

Obtaining and Making Ready the Skull

We usually use bovine skulls and, where possible, we seek to obtain skulls from mature animals, preferably from a cow, rather than a bull. We are also able to obtain a few sheep skulls each year. Josephine Porter also used goat skulls to good effect. Our experiment with a pig skull proved to be somewhat unsatisfactory since the slaughterhouse procedure for a pig skull involves sawing the skull and splitting the brain cavity into two separate pieces. Efforts to wire the skull back together and produce an acceptable preparation were less than successful.

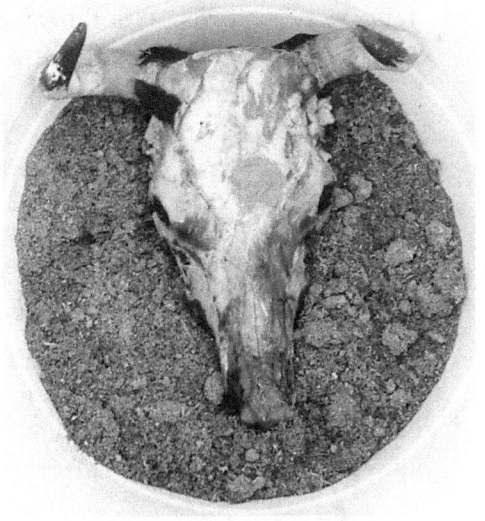

The untrimmed cow skull is placed in a barrel before being covered by moistened sawdust. (photo © Patricia Smith)

The skulls we use are fresh or, at least, fresh frozen. The source from which we obtain the skulls has already removed the brain shortly after slaughter and removed the skin from the skull as well. Since we sometimes put down as many as sixty skulls at a time, we have the forepart of the head sawn off as well, so that the site where the skulls are buried does not have to be enlarged to accommodate skulls with a full snout on them. The saw cut is usually made ahead of the eyes, but behind the innermost teeth, and the lower jawbone is also sawn off close to the hinge at the base of the skull. When we receive the skull, there is still a considerable amount of flesh and fat as well as the eyes, all of which we prefer to remove before we stuff the oak bark into the brain cavity.

Barrel Method Using Sawdust

We are also experimenting with another procedure where the brain cavity is stuffed with the oak bark without trimming. The now stuffed but untrimmed skull is then exposed to flies, which lay their eggs in the still untrimmed fleshy parts. After a day or so of such exposure, the skull is placed in a barrel filled with well-moistened sawdust. After a fairly long time, often many weeks later, the skull is removed from the barrel, and if the fly maggots have done their job, the flesh and eyes should have been consumed.

The entire time that the skulls are residing in the barrel, one needs to ensure that fresh rainwater is added to the barrel periodically. To date, we have not been completely satisfied with this technique, especially when dealing with a large number of skulls, but we intend to experiment further with it.

By controlling the amount of water pressure, this homemade tool of copper tubing and a hose fitting with a shutoff valve is used for flushing out the brain material and leaving the membrane lining of the skull intact. This type of tool is also used in slaughterhouses for the same purpose. (illustration © Mattias Baker)

505: How to Make the Oak Bark Preparation

For our own animals, when we do an on-farm slaughter, we use a special tool to remove the brain from the cavity. This tool consists of a length of copper tubing (one-quarter inch diameter), which has been soldered to a hose fitting that includes a shutoff valve. When removing the brain, the tool is affixed to the end of a water hose. The tube is inserted into the brain cavity, and the valve is turned on. The brain is then flushed out of the cavity by the high pressure of the water, but the membrane lining the skull cavity is not damaged by the brief exposure to the water. When we are dealing with frozen skulls, we employ this tool again when the skull has thawed to ensure that all of the brain matter is out of the skull cavity. This is the same tool and technique used in the slaughterhouse for removing the brain from the skulls. In the case of the skull from our on-farm slaughter, we proceed to completely remove the jawbone rather than sawing it off close to the skull as the slaughterhouse does.

Stuffing the Brain Cavity

Once the skulls are ready for stuffing, we take the moistened oak bark, and initially, at least, with the help of a small funnel with a wide opening, we begin to insert the oak bark into the brain cavity through the foramen magnum, the opening at the base of the skull. We try to pack the moistened oak bark as tightly into the cavity as we possibly can, using a dowel to help with the packing. When we believe the skull is fully packed, it is placed in a tub of water. If properly and tightly packed, the skull should sink, or barely float.

If the skull floats high out of the water, one should redouble the effort to pack it even more tightly, because there is a very great likelihood that there is either an air pocket or worse, some brain tissue remaining within the skull cavity. When the skulls are all fully packed, the *foramen magnum* opening is sealed with clay, and if possible, a piece of bone is cut to fit the opening. When dealing with forty or fifty skulls, or even a dozen, it is sometimes quite difficult to find enough bone pieces to use for this step.

A narrow path of rainwater trickling through a shallow gully makes for a suitable site for burying the 505 in the fall. (photo © Patricia Smith)

Choosing the Site to Bury the Skulls

If one does not use the rain barrel technique suggested by Steiner,[3] then a choice of a low swamp-like spot preferably isolated from any other ponds or waterways could be utilized. It should also be a spot where rainwater flows or collects. The main site we have chosen has proved to be a very good one over the years. We generally make and

---

3   Ibid., p. 101.

bury the preparation in the fall of the year after Michaelmas (Sept. 29), but before Thanksgiving (the fourth Thursday in November). After the skulls are placed in the pit, we make sure that rainwater can be channeled into it, and we cover the skulls with a combination of mud and vegetable matter, mostly leaves from the forest floor around the pit, since it is much handier than would be the case were we to use peat moss, as Steiner suggests.[4]

To prevent stray dogs or a fox from digging up the skulls, we cover the site with a piece of woven wire fencing, which is then staked down at the edges with metal stakes. Large rocks, pieces of metal pipe, or old metal fence posts are also placed over the fencing to hold it down in place to thwart any animals seeking to dig anything up. Josephine Porter chose to encircle her oak bark pit site with a woven wire fence requiring several wooden posts. Failure to protect the site can lead to a major loss when it comes to recovering the preparation.

Recovering the Oak Bark Preparation

When it comes to digging up the preparation, the time of excavation would normally be somewhere between Easter and St. John's Day (June 24), with a time at or after Ascension Day (forty days after Easter) preferred. However, because we have for the last several years conducted seminars here at JPI in preparation making, we have chosen to delay the digging up of this preparation until the time of our Fall Seminar (usually the third week of October) so that the participants can experience both the finished preparation and the making of it as well. In other words, they can experience the preparation from start to finish. Such a delay does not seem to diminish the potency of the oak bark preparation.

After the skulls are dug up from the pit, they are all carefully hosed off, so that we can avoid contaminating the finished preparation with the debris from the mud and vegetation. The finished preparation is

---

4   Ibid.

then carefully removed from the brain cavity using a "nit-picker" or a piece of stiff wire that has had a hook bent into one end of it.

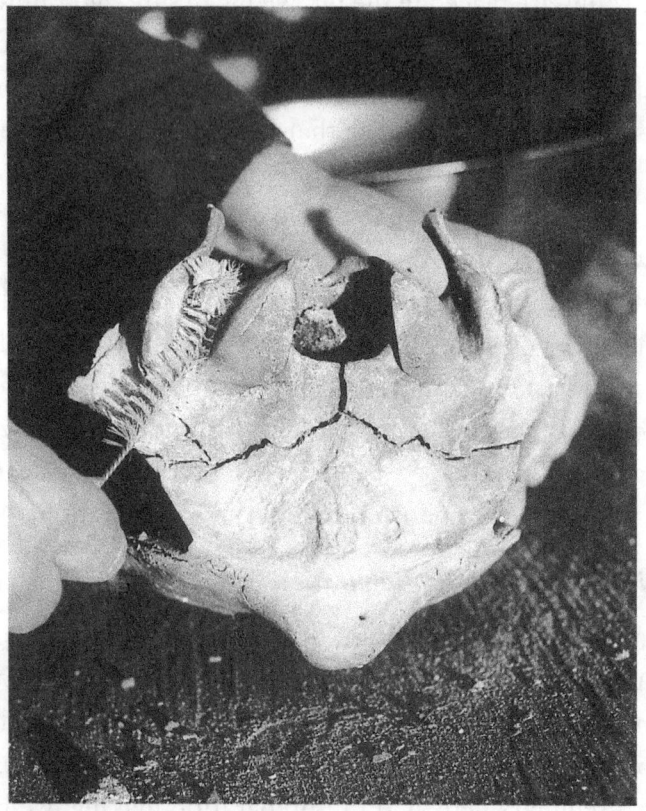

Once excavated, the skull is gently hosed down to remove mud and debris and is left to dry for a few days. After drying, fine grit and sand is further removed from the skull and around the opening with a bristle brush before attempting to empty the finished oak bark preparation within the skull. (photo © Patricia Smith)

The finished preparation is carefully emptied into a bowl or other container. When sufficient material has been removed from the skull cavity, and there is still oak bark that cannot be reached with whatever tool we are using, we then turn the skull over and tap the back of it vigorously with a hammer so that the remaining oak bark is

505: How to Make the Oak Bark Preparation

loosened and exits through the foramen magnum into a container. When all the skulls have been emptied, the preparation is screened through one-eighth-inch mesh hardware cloth to remove any foreign matter such as bits of skull membrane, bone, stones, soil, vegetation, etc. We will often find earthworms of a most peculiar nature, which we also remove. The finished preparation is then placed in a fifteen-gallon crock in our root cellar and is ready for use.

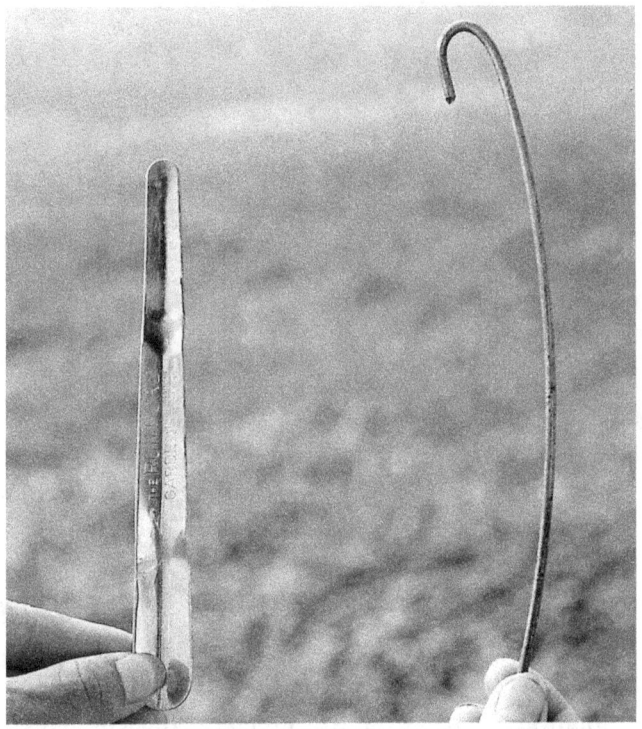

Tools for removing the finished oak bark from the skull cavity: a "nit-picker" on the left and a stiff wire hook (photo © Patricia Smith)

Our volumetric measure for a unit of oak bark is a rounded teaspoon, which is about two grams by weight. Since most farmers and gardeners do not have a scale in grams ready to hand, we use the volume measure as more convenient for JPI's clients. This quantity, along with the other five compost preparations is sufficient to treat ten to fifteen tons of compost material. When one builds the initial compost pile on a site, the recommendation is to calculate one set for every ten tons of material. As subsequent compost piles are built on the same site, the ratio of compost sets to the material can be adjusted until one set treats as much as fifteen tons of material. For those who calculate compost material by the cubic yard, the conversion factor is one and one-third cubic yards (a cube 4' x 4' x 4') equals one ton.

Emptying the finished oak bark from a skull on the left yields about one cup of finished preparation. On the right is the debris cleaned from the skull. (photo © JPI staff)

For use by itself, the oak bark preparation can be employed on soils where the calcium levels are too low, but it is probable that repeat applications would be required in order to significantly affect available calcium levels. As a prophylactic measure against plant diseases, a separate application of the oak bark preparation should also be considered before planting any crops. For use as a seed bath, the oak bark preparation is extremely beneficial for oats, bush beans, and especially lettuces. Germination is much faster, and the overall health of crops is improved.

Other uses for this preparation would suggest themselves by focusing on what Steiner has to say about its purposes. In introducing this preparation, he speaks of it as a universal remedy against

505: How to Make the Oak Bark Preparation

plant diseases.[5] He also speaks of the need to bring calcium that remains within the realm of the living to the soil.[6] Such calcium in a living state can restore "order when the ether body is working too strongly"[7] and thereby prevents the astral from gaining "access to the organic entity."[8] Calcium in this form "damps down the ether-body, and thereby makes free the influences of the astral body."[9] By thinking through such descriptions, we may conclude that, in a very wet season, when there is quite prolific or "rampant" growth, a separate application of the oak bark preparation might tame such growth and allow the fruiting and flowering process to proceed in a more normal fashion.

One anecdotal experience from a gardener in Pennsylvania involved using the oak bark preparation in combination with the stinging nettle preparation (BD504) as a "watering in" treatment for tomato plants to restore the leaf growth of those plants when they had been over-stimulated toward flowering and fruiting by repeated diligent applications of the silica (BD501) and valerian (BD507) preparations. These applications of BD501 and BD507 did improve the fruiting, but it left the tomato plants almost barren of leaves and considerably less photogenic than was needed for a newspaper photoshoot scheduled within the following two weeks. In another observation, the same combination of oak bark and stinging nettle also seemed to have a modest ability to thwart frost damage when used late in the fall of the year to water in pepper plants.

For the separate application of the oak bark preparation as a soil or plant spray, a stirring time of one hour is preferred, although for small areas or for seed baths, one may consider a potentization in an apple cider jug, which is shaken in a lemniscate motion (see Issue No.

---

5 Steiner, *Agriculture*, p. 100.
6 Ibid., p. 101.
7 Steiner, *Agriculture Course* (UK ed.), p, 97.
8 Ibid.
9 Ibid.

28 of *Applied Biodynamics* for a more detailed description of this process). Above all, we strongly encourage you to experiment with the use of this preparation on its own and to share the results of those experiments with the readership of *Applied Biodynamics*.

The Oak Bark Preparation:
Organ of Living Thinking (in the Farm Organism)

Steiner specifies merely the skull of a domestic animal[10] as the sheath for the oak bark preparation. Indeed, he does not even define for us what a domestic animal is, and at times, strangely enough, the question has come up from the literalists amongst us regarding the suitability of a dog or a cat skull, since they too are generally regarded as domestic animals. Lily Kolisko seems to suggest[11] that using the skull of a horse might yield a poorer end product. It has been suggested that in a certain sense, the cat, dog, and horse are somewhat too close to the human being and should be avoided. In order of availability and my personal preference, I would rank the choice as follows: cow, steer, heifer, and bull among the bovine species; then sheep, goat, and pig in that order, with both sheep and goat preferred over a bovine bull skull. We have had so little opportunity as far as using a pig skull that we can offer no firm opinion as to its suitability. While we have experimented with using a buffalo (American bison) cow skull, since it is generally not regarded as a domestic animal, the resultant oak bark has not been subsequently used and awaits the leisure of future experimentation. No opportunity has presented itself to try more exotic animals such as llama, alpaca, or water buffalo.

The most important single factor in using any skull is to ensure that it is as fresh as possible, so that the membrane lining the skull cavity, the meninges, is intact. In a certain sense, the true sheath for the oak bark is not the skull, but rather the meninges. The practice of using a desiccated skull, either by placing the freshly slaughtered

---

10 Steiner, *Agriculture*, p. 101
11 Kolisko, *Agriculture of Tomorrow*, p. 183.

505: How to Make the Oak Bark Preparation

skull in a compost pile or the crotch of a tree for several weeks or months, can only lead to a finished preparation of highly questionable quality and ranks very high as a shortcoming in our making of this preparation. To remain within the realm of the living as Steiner encourages us to do[12] can only be accomplished for this preparation by preserving the true sheath of the meninges membrane that lines the brain cavity. If we are too finicky or averse to using a somewhat gory fresh skull because we don't want to have to deal with something quite that messy, and allow a compost pile or the weathering of nature to clean up the skull for us, then in all likelihood, the meninges will have also been cleaned up as well. We must not allow our sentimentality or aversion to the unpleasantness of the task to deter us from making a vital and usable preparation. I would venture to say that over the years, quite a lot of substandard oak bark preparation has been produced. By so doing, we have not brought to our biodynamic practice the genuine disease countering forces that Steiner offered when he described this preparation.[13]

I would have to say that if a practitioner can't get over their concerns and objections to using a fresh skull, then they should not claim to be a maker of the biodynamic preparations. They are instead a mere dilettante just going through the motions. There have been some who have presented as an excuse for using a desiccated skull the fact that the end product when using a fresh skull had a very rank and objectionable odor. My response to anyone who ends up with a rotten smell when they empty an oak bark skull is that they most likely had neglected to remove all of the brain tissue when they first prepared the skull for making the preparation. If one is thorough and not sloppy in their preparation making, they will be a lot further ahead. Doubts have also been voiced recently concerning the practice of harvesting oak bark,[14] with the suggestion that the outermost layer of

---

12 Steiner, *Agriculture*, p. 101.
13 Ibid., p. 100.
14 Malcolm Gardner, "Are We Collecting the Best Oak Bark?"

bark is not what should be harvested, but rather bark next to the cambium layer, and from younger trees and limbs in the spring of the year as *was/is* the recommended practice when using oak bark for human or animal medicaments. To address these doubts requires a certain historical perspective, as well as a genuine look at what Steiner is seeking to have us create in this preparation.

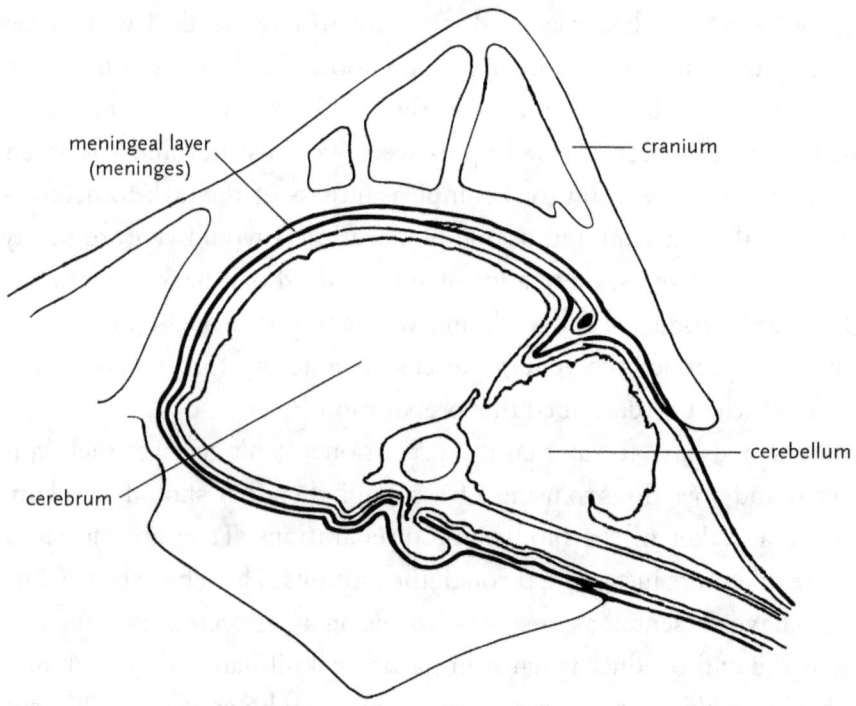

The membrane lining within the skull, the meninges, surrounds the cerebrum and cerebellum. Original illustration before modification (Pasquini, Atlas of Bovine Anatomy, p. 86)

When the biodynamic preparations first began to be experimented with, even before Rudolf Steiner gave the agriculture lectures, several people were involved in following his instructions, including Lily Kolisko, Ehrenfried Pfeiffer, Günther Wachsmuth, and Immanuel

505: How to Make the Oak Bark Preparation

Voegele. My instructions for making the oak bark came from my mentor, Josephine Porter, who learned her preparation making from Pfeiffer. She was most specific when teaching me to harvest the bark from the white oak tree—saying that the outermost layers, excluding the cambium layer especially, were the proper choice to harvest for this preparation.

Many thread-like earthworms such as this are found in the finished oak bark preparation at JPI. (photo © Patricia Smith)

Likewise, my recent conversations with Ruth Zinniker—who learned to make her preparations from her father, Immanuel Voegele—confirmed that she, too, was taught to harvest the outermost layers of the oak bark, in her case from *Quercus robur*, the English or common oak, as it is sometimes referred to in Europe. Since we both received explicit instruction from people either directly connected with Steiner, or only one person removed, I think it is important to approach a different understanding of just why one would use this particular bark, rather than approaching the oak bark harvest with the same mindset as though one were harvesting to make a medicament for humans or animals. We should not presume that the biodynamic world has been harvesting the wrong kind of bark all these years simply because this differs from the way one harvests this substance for making a medicament.

Left: The bark of a white oak, Quercus alba (photo © Patricia Smith)
Right: Harvesting oak bark (photo © JPI staff)

In the first place, to draw an analogy from the steps used in making medicaments for human or even animal use is to assume without really examining what Steiner is seeking for us to achieve with the oak bark preparation. Of the several preparation plants, all of them have been used as medicaments for long periods of human history. However, it is of more than passing interest that Steiner recommends in more than one instance, parts of these plants that are not normally used in making remedies for treating humans and animals. In the case of valerian, Steiner suggests the flowers for use in making an agricultural preparation, when for humans the recommended part of the plant is the root. This is true, as well, for the dandelion, where it is the flower we seek for biodynamic application rather than the root, which is the choice when dealing with human remedies. In a reference from my reading of Steiner, I experienced a moment of awareness and recognition when he refers to the bark production of a tree as a kind of "flowering process." Unfortunately, I cannot cite that reference for the reader's confirmation, and I do not have the leisure to re-read the several lecture series in which I may have seen such a statement. By recognizing that the outermost layer of oak bark has an inner quality or nature of a flower (tending toward seed?),[15] all of the compost preparation plants, including stinging nettle, are thereby to be seen as encompassing this flowering process. It is at the flowering stage that the astral most closely approaches the plant.[16]

Carl König's Observations

So that you may perceive that I am not alone in my conclusions, Karl König in the very valuable book *Earth and Man*, uses the following words when he examines this oak bark preparation:

> "...the oak bark preparation, placed into the compost heap, makes the compost heap not only sensible but conscious, which is much more.... What makes the animal conscious, gives to man conscience and not merely consciousness. The same powers give plant health. A healthy plant is healthy by virtue of the same powers that create consciousness in an animal and build up conscience in man." (p. 308)

Is it perhaps the bark of a tree that ensures the hovering astrality that Steiner refers to in the *Agriculture Course*? When Steiner indicated the oak bark preparation, the goal was to have a substance so treated that it had the capability of damping down the etheric.[17] Such capability comes about by stimulating strong astral energy. Within the nerve-sense organism of the human being, the astral dominates, as well. It is within that nerve-sense pole that Steiner speaks of the need for a state of nearly deathlike existence for the nerve cells, to bring about the possibility of human thinking. To create an organ of thinking, as it were, for the compost pile, and ultimately for the earth and soil treated with that compost, Steiner suggests for us a substance so closely approaching a state of deadness, that it can be

---

15 Steiner, *Agriculture Course* (UK ed.), p. 92.
16 Steiner, *Agriculture*, p. 153.
17 Ibid., p. 97.

seen as the nearest still living substance within the plant kingdom akin to the brain cells. This substance, the outermost layer or rind of bark from the oak tree, is then placed within the meninges sheath guarded by the skull of a domestic animal. The subsequent burial in a watery pit during the winter months when the Earth is most inwardly alive is what allows the creation of a brain-like organ for first the compost, and then the soil.

Just as the brain is surrounded by the fluid in which it is floating, so, too, does a watery earthy fluid surround the oak bark within the meninges that resides inside the skull. It is only this step of creating an organ of thinking, or consciousness, added to the other organs created by the several compost preparations, that allow the possibility for a farm organism to develop and to become an individuality. It is only when a farm individuality or a farm *I* can manifest that we achieve the purpose of the biodynamic preparations "so that the earth may be healed."[18]

---

18 Ibid., p. 7.

506

Taraxacum officinale
Dandelion

+

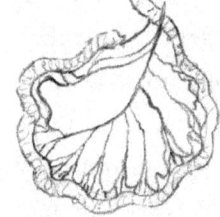

bovine mesentery

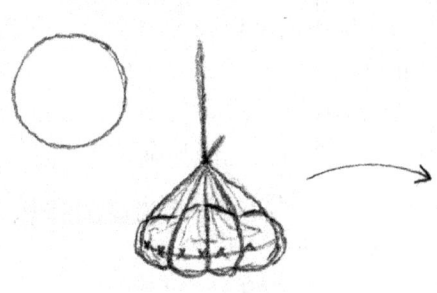

506: Dandelion: Messenger of Heaven

Abigail Porter
Applied Biodynamics, *no. 76, 2012*

What could be finer than a glorious field of glowing dandelions in full bloom on a beautiful, clear, spring day? Writing this in the middle of winter, I have to smile at the image of all of those happy little "sunshines" brightening the day. I always knew that dandelion blossom picking season was fast approaching when my mother brought that first bowl of dandelion greens, topped with the warm sweet and sour dressing to the table as a spring tonic to chase away the winter doldrums. Dandelions, at one time, held a noble position in the world. Not only were they highly valued for their medicinal qualities and culinary uses, they were also valued for their beauty. Dandelions were deliberately planted in European flower gardens and were so dearly loved that poems were written to honor them. Children have loved to play with them, making chains, garlands, and crowns and blowing the fluffy seed head to tell the time, make wishes, or tell fortunes. In Japan, whole horticultural societies were formed to admire them and develop new varieties.[1]

People used to pull the grass out of their yards to allow dandelions to grow. All parts of the dandelion—root, leaf and blossom—are used as food and for medicine. It is a wonderful detoxifier in our bodies as well as in the soil where it is growing. A fertilizing tea can be made from the whole plant. The roots or blossoms can be used to make dyes and the leaves and blossoms can be made into paper. The

---

[1] Sanchez, *The Teeth of the Lion: The Story of the Beloved and Despised Dandelion.*

dandelion plant produces ethylene gas and can be placed in a bag with green fruit to promote ripening. During World War II, dandelions were even made into latex rubber. Dandelions provide food for the soil and for animals, birds, bees and butterflies. What an amazing and versatile plant for the healing of the earth!

History of Dandelion

The common dandelion or *Taraxacum officinale* is one of thirty-four macrospecies and seventy to two thousand microspecies (depending on who is counting) of the genus *Taraxacum* belonging to the *Asteraceae* (*Compositae*) family. It is related to both sunflowers and lettuce. Taraxacum is thought to have originated in Eurasia, maybe in the Himalayas, and is widely dispersed throughout the world, being more prevalent in the Northern Hemisphere. *Taraxacum officinale* was repeatedly brought to the New World as early as the mid-1600s.

Colonists are credited with introducing dandelion to the Native Americans, but it is much more likely that they were already using one of the three or four native species—perhaps the alpine or horned dandelion (*Taraxacum ceretophorum*), which grows in New England as well as on the west coast and most of Canada. Dandelion has a long history on this continent; in Alaska there are dandelion fossils over 100,000 years old.[2]

Botany

Dandelion is a perennial with a long taproot and leaves that are usually sharply toothed (in the wild variety) and shaped to funnel dew and rain down to the crown and root. It is one of the first greens to appear in early spring and the blossoms, being very sensitive to sunlight, open in the morning and close toward evening with the decreasing light. They will not even open on cloudy or rainy days. Being a *Compositae*, what appears as a single flower is

---

[2] See watchingtheworldwakeup.blogspot.com/2008/04/dandelions-are-way-cool-part-3-where.html.

really many small flowers or florets, each one a valuable source of nectar and pollen for the beloved honeybee and other insects. Dandelions reproduce either by pollination or asexually or both. Dandelions can self-pollinate, or wind and insects can spread the pollen, but sexual reproduction is rare. *Taraxacum officinale* reproduces primarily asexually through seeds by apomixis or agamospermy, generating clones of the mother plant without pollination. Pieces of the root will also grow new plants. Dandelions bloom all through the year, with a major flush in the spring and a lesser one in the fall. Each blossom can produce fifty-four to 172 seeds, and each plant can produce five thousand seeds during the course of the year. With the right wind currents, the seeds and their fluffy parachutes can be carried a hundred miles. It is no wonder that dandelions have been associated with Jupiter, the planet of fertility, abundance and expansion.

Illustration by Doris Clark (reprinted with permission from Ehrenfried Pfeiffer, Weeds and What They Tell Us*)*

A Dynamic Weed

Dandelion proliferates where the earth has been "disturbed" or where humans have cultivated the soil. On this continent (North America), dandelions were still valued until the mid-1900s when the petrochemical companies began their massive advertising campaigns for the eradication of dandelion.

Some cities even passed laws making it illegal to cultivate dandelion within the city limits. Although not up there in the much-maligned ranks of kudzu, Japanese knotweed, or wild rose, dandelion (*Taraxacum officinale*) is considered to be invasive to the North American continent since it is not native and grows abundantly where it is not wanted. In his book, *Invasive Plant Medicine*, Timothy Lee Scott proposes that invasives proliferate where the native ecology has been compromised by pollution or by man. The invasives serve to heal the damage to the earth. Once the soil is brought back into balance, other plants flourish and the invasives move on. Dandelions can detoxify soil contaminated with heavy metals and are used in phytoremediation programs.

Dandelions fertilize depleted soils with nitrogen, potassium, calcium and with trace minerals brought up by their deep taproot, bringing soils back to life. They can live for five years, and their taproots can go deep and break through hardpan, aerate the soil, and reduce erosion. E. E. Pfeiffer lists the dandelion as a "dynamic" weed (a plant that changes the soil or plants around it in a special way). Dandelion provides humus and minerals, restoring what has been lost or washed away. Pfeiffer writes that the dandelion is the plant equivalent of the earthworm.[3]

The benefits provided by dandelion are analogous to those provided by earthworms, creating humus, aerating the soil and transporting nutrients. Earthworms favor the soil around dandelions, and adding the plant to the compost pile will attract them to the pile.

Dandelion as Medicine

Timothy Scott relates the epidemic rise in diseases such as diabetes, cancer, heart disease, and Lyme disease to the spread of invasives, which are healers for these same diseases. Nature is providing what is needed. Dandelion has a chelating capacity and detoxifies

---

3   Pfeiffer, *Weeds and What They Tell Us*, p. 72.

the human body of heavy metals and purifies the blood, liver, and kidneys. It is antibacterial, antiviral, antiparasitic, anticarcinogenic and antiinflammatory. It has been used for heart disease, cancer, diabetes, obesity, hepatitis, and depression. Rich in antioxidants, vitamins, and trace minerals, it helps to rebuild the body. More than fifty percent of the herbal blends on the market contain dandelion, attesting to its wide range of healing benefits. "We are fortunate to have the wisdom of nature... to counterbalance the foolishness of humans."[4]

Over the centuries, dandelion has been prescribed to increase the milk flow in nursing mothers and dandelion has been fed to cows to increase cream and milk production, its effectiveness witnessed by Scott's neighboring farmer.

Dandelion has been used as medicine for thousands of years and has so many health benefits that it has been called "the poor man's ginseng." If it were as scarce as ginseng, it would most likely be as costly. One species or another has been used in Traditional Chinese Medicine, Greco Arabic, Ayurvedic, homeopathic and western botanical medicine. In fact, the Arabic translation of the word *taraxacum* is "remedy for disorders."

The root is considered to be the most potent part of the plant and is used for conditions of "heat" and inflammation, particularly in the liver, gall, kidney, digestive tract, and skin. It is also recommended for tumors, abscesses, and pustules. The leaves are diuretic and as effective as conventional drugs yet don't have the side effect of potassium depletion, as they are rich in this mineral.

The French word *Pissenlit* (piss-in-bed) is one of dandelion's many common names and reflects one of its properties. A tea from the leaves has also been used to treat bladder infections as well as depression. A tea from the blossoms is said to alleviate pain. And from the 1500s, *A New Herbal, the Historie of Plants,* by Rembert

---

[4] Scott, *Invasive Plant Medicine*, p. 100.

Dodoens, notes, "...bruised and eaten with his leaves and roots, is very good against the bitinges of venomous Serpentes."[5]

> "Dandelion.... It is a plant with hundreds of secrets.... You will not find anything more expressive of the workings of silica."
> —KARL KÖNIG[6]

Nicholas Culpeper, eighteenth-century English botanist, herbalist, physician, and astrologer might have been the last well-known physician to practice medical astrology (Iatromathematics). Culpeper places dandelion under the dominion of Jupiter for its opening and cleansing qualities. Combined with Alisanders (Alexander or wild parsley), it is a "wonderful help" for "whoever is drawing toward a consumption, or an evil disposition of the whole body, called cachexia..."[7]

Jupiter

In ancient times, astronomy, astrology and medicine were closely related to each other, and it wasn't until the eighteenth century that they became distinct fields of study. Dandelion has a long history of being associated with Jupiter and epitomizes the qualities for which Jupiter is known: abundance, growth, freedom, fertility, wisdom and jovial exuberance.

Bitter herbs are healing for the liver, and those especially with yellow blossoms are under Jupiter's domain. Jupiter is the largest planet and is associated with the liver, the largest internal organ, and with the blood. Jupiter's signs are Sagittarius (diurnal) and Pisces (nocturnal), which are opposed to Mercury with Gemini and Virgo. Tin is Jupiter's metal, and the word tin comes from the Etruscan god

---

5 Dodoens, *A New Herbal, or Historie of Plants*, p. 565.
6 König, *Earth and Man*, p. 304.
7 See www.complete-herbal.com/culpepper/dandelion.htm. Wasting of body mass that cannot be reversed nutritionally: en.wikipedia.org/wiki/Cachexia.

Tinia, who was equivalent to the Norse god Thor, the Greek god Zeus, and the Roman god Jupiter or Jove, god of the sky and lightning, and king of the other gods. Jupiter's mass is 2.5 times that of all the other planets in our solar system combined. Its magnetic field is 20,000 times stronger than Earth's, and its fast rotation creates electrical storms.

A Jovial day is only about ten Earth hours but a Jovial year takes twelve Earth years. Jupiter spends one year in each constellation of the zodiac. Jupiter's composition, which is more like a star, is similar to the sun. It is composed of mostly hydrogen (approximately ninety percent) and helium (ten percent) by volume, and, like the sun, it has its own kind of solar system with many moons orbiting it. Unlike the sun, its surface temperature is very cold.

Dandelion in the Kitchen

An entire book could be written about the culinary uses of dandelion. It is one of the most nutritious plants that we can add to our menu. It is rich in antioxidants, trace minerals, calcium, potassium, protein, and vitamins B, C, D and K. It has more beta-carotene than carrots and is the third richest source of vitamin A, following cod liver oil and beef liver. "Gerasimova, the Russian chemist who analyzed the dandelion for, among other things, trace minerals, stated that 'dandelion [is] an example of a harmonious combination of trace elements, vitamins and other biologically active substances in ratios optimal for a human organism.' (Hobbs 1985)."[8]

Beyond the more common uses of the leaves in salads, the root as a coffee substitute and the blossoms for wine, the greens can be cooked and used in the same way as spinach. All parts of the plant can be stir-fried or used in soups. The crowns are considered a delicacy when sautéed in plenty of butter and a squeeze of lemon. The roots combined with burdock have been made into traditional

---

8 Peter A. Gail, "Health Benefits of Dandelions," www.leaflady.org/health_benefits_of_dandelions.htm.

therapeutic British beer and soda. The blossoms can be made into fritters, syrup and jelly. Dandelion is the perfect survival food and free health insurance.

I hope this inspires you to bring more dandelions into your life. With the increasing interest in using food as medicine and with the growth of biodynamics, maybe Dandelion will again have its "Day(s) in the Sun."

Abby Porter grew up on a biodynamic dairy farm in Northeast Pennsylvania. Following a career in the arts (jewelry design and production), she returned to her roots with a move to Berkeley Springs, West Virgina, and felt a need to rehabilitate the poorest garden soil she had ever seen. Abby served on the Board of Directors of the Josephine Porter Institute for Applied Biodynamics for five years and wrote for their periodical, Applied Biodynamics, *for ten years. She currently consults and leads workshops on beginning biodynamics for backyard gardeners.*

506: Dandelion Compost Preparation

Abigail Porter interviews Hugh J. Courtney
Applied Biodynamics, *no. 76, 2012*

The dandelion blossoms are an essential ingredient in the biodynamic compost preparations. In the following interview, Hugh Courtney explains the value of dandelion and the role that the dandelion preparation BD506 plays in the compost pile, in the soil, and through plants.

> *Abigail Porter: In lecture five of the Agriculture Course, Rudolf Steiner chooses very specific plant materials to vitalize the compost pile, which in turn enlivens the soil and plants. What is the significance of Steiner's choice of dandelion for one of the compost preparations?*

Hugh Courtney: Steiner chose plants that were readily available to the farmers attending the course and the ones that most strongly exemplified the qualities that he wanted to bring to the soil to enliven it. Steiner referred to the dandelion as the "messenger of Heaven." It has the ability to attract silicic acid[1] from the whole earthly and cosmic environment. And it is dandelion's relationship to potassium and the silica process that is needed.

---

1 Silicic acid is a general name for a family of chemical compounds of the elements silicon, hydrogen, and oxygen with the general formula $[SiO_x(OH)_{4-2x}]n$. Silica is the chemical compound silicon dioxide, an oxide of silicon with the chemical formula SiO_2. Silicon is a chemical element with the symbol Si. Silicon is the eighth most common element in the universe by mass, but very rarely occurs as the pure free element in nature. It is most widely distributed in dusts, sands, planetoids, and planets as various forms of silicon dioxide (silica) or silicates. Over ninety percent of the Earth's crust is composed of silicate minerals, making silicon the second most abundant element in the Earth's crust (about twenty eight percent by mass) after oxygen. Silicon is an essential element in biology, although only tiny traces of it appear to be required by animals. It is much more important to the metabolism of plants.

506: Dandelion Compost Preparation

It is hard to speak of one preparation without relating it to the others.

To understand dandelion, one of the things you need to do is look at the preparations in terms of polarity. Dandelion preparation BD506 is in polarity to chamomile preparation BD503. Dandelion relates to Jupiter and the silica process. By contrast, the chamomile relates very much to Mercury and the limestone process. These are *processes,* and not what we know as the elements on the periodic table or chemical compounds.

Silicic acid and the silica process are needed as a kind of cosmic catalyst for many processes in both the plant and animal kingdom. The dandelion plant attracts silicic acid and mediates between the silicic acid in the Cosmos and the silicic acid that is needed in the earthly realm. Dandelion stimulates this cosmic nutritional stream. By burying dandelion blossoms in the peritoneum through the winter, exposed to the winter Earth forces, these qualities are enhanced or potentized so they can be transferred to the compost pile, the soil, and the plants. It gives the soil and the plants the ability to attract as much silicic acid from the atmosphere and the cosmos as is needed. It allows the plant to become sentient, more sensitive, and aware of its surroundings, and it will draw to itself whatever it needs, both substances and forces, from near and far, earthly and cosmic.

"The innocent yellow dandelion! In what ever district it grows, it is the greatest boon.... [BD506] will give the soil the faculty to attract just as much silicic acid from the atmosphere and from the Cosmos as the plants need, to make them really sentient to all that is at work in their environment. For they of themselves will then attract what they need."
—RUDOLF STEINER (*Agriculture Course*)

*AP: Steiner uses both terms—*mesentery *and* peritoneum—*when referring to the sheath for the dandelion. Are they the same thing?*

HC: Yes and no. It is the same membrane but in a different place. The peritoneum is the membrane sack or pouch that holds everything in the abdominal cavity: the four stomachs, the intestines, liver, spleen, gallbladder, etc. If you follow the peritoneal tissue toward the spine, that same tissue folds into the cavity creating a sleeve of sorts, which holds all of the intestines together and supports them. This part of the peritoneal tissue is called the mesentery, or "isles of mesenteria." In another place the peritoneum folds into itself and creates a mesenteric pouch, or sling, to support the abomasum, the fourth and last stomach, also called the true stomach. It is also referred to as the "net" stomach, perhaps because of this peritoneal membrane supporting it. In the discussion after lecture five in the Agriculture Course, Steiner uses the term *Bauchfell*, which translates to the peritoneum. He also uses the word *Gekrose*, which translates to the mesentery membrane that surrounds the intestines. At JPI we use both the peritoneum and the mesentery. We have found that a more humus-like preparation is obtained when we use the peritoneum as long as the inside of this tissue is used next to the dandelions, that is, the side facing the interior organs, not the side facing the outside or skin of the abdominal cavity. You have to make sure that the dandelion is in contact with the interior side of the peritoneal membrane. If one places the dandelions on the outside of this tissue, the result will not be as transformed and

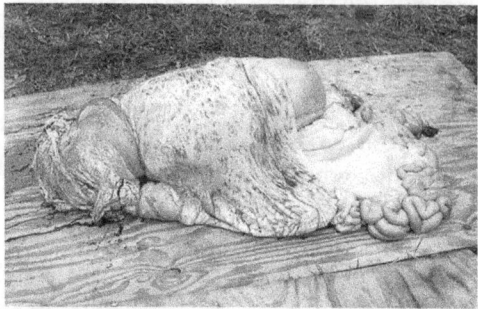

Top: *Entrails from a cow. Peritoneal tissue to the left lying on stomachs, and the "Isles of Mesenteria" holding the intestines together on the right.*

Bottom: *Peritoneum tissue*

506: Dandelion Compost Preparation

can be foul-smelling and slimy, with the structure and color of the original blossom still identifiable. At slaughter, we label one side to make sure we use the right side. It doesn't seem to matter which side of the mesentery is used, but as I said previously, a more humus-like preparation results from using the peritoneum.

AP: If one must obtain the peritoneal material from a slaughterhouse and cannot be present during slaughter, how would one tell which is the inside, which needs to be placed next to the dandelions?

HC: It would be good to have the butcher mark the inside of the tissue for you at the time of slaughter. If the peritoneum has been handled carefully and is not dirty, one can observe that the inside is slightly smoother than the outside.

AP: And what is the importance of the peritoneum or mesentery in biodynamics?

HC: The value of the peritoneum or mesentery, which is filled with nerves, blood vessels, and lymph vessels, is that it is the inner world. That inner world of the abdomen is totally connected to the cosmos and the cosmic nutrition stream. It has no connection to the outer world. This is in contrast to the intestines, which are connected to the outer or earthly realm through the mouth and the anus. The intestines and stomach are connected to the external world and the earthly nutrition stream. The other organs within the peritoneal sheath– the liver, kidneys, and spleen, including the mesentery–are closed off from the external world. It is important to make sure that the dandelion is in contact with the interior surface of the peritoneal tissue so that both the dandelion and the peritoneum are all still an inner process, still connected to the inner world of the cosmos.

AP: What is the function of the dandelion preparation in the compost pile?

HC: It brings the silica process and the liver process to the organism of the compost pile and subsequently to the farm entity. With the compost preparations, you create a living entity with its own organs. The preparations radiate throughout all the material in the pile and help organize the pile. You can relate this invisible radiation process to the rhythmic system, the blood, and the breathing process.

If you look at what Steiner says early in the Agriculture lectures, he talks about the agricultural individuality and that this individuality, the farm, should be seen as an upside-down human being, with the head in the soil. Growing above the soil is the plant and at the soil surface is the area that he relates to the human diaphragm.

Both a bit above the soil and a bit below the soil you have a kind of digestive process going on. With the preparations, he gave us the keys to developing the organs of the farm individuality, so in essence one of the biodynamic farmer's goals is to establish the farm with an "I" as it were. That happens through the use of the preparations, and each of them brings a certain organ process, related to the human being, to the compost pile, and ultimately to the farm individuality. The skull and oak bark bring in the brain and consciousness, which is essentially a Moon process. The nettle relates to the heart and spleen and the Mars process and allows the compost and soil to be more feeling, more sensitive, sentient, and intelligent. The yarrow and bladder relate to the lungs and kidneys and the Venus process, while the chamomile and intestines relate to the full digestive process and as referenced earlier, the Mercury process.

With valerian, you have the Saturn process, the framework of an organism; the skeleton, the bones, and the skin. I can see why that might be the reason Pfeiffer put the valerian preparation in the pile as well as sprinkled it on the surface of the pile. The dandelion preparation, in particular, brings the liver process, which cleanses and operates through silica. It brings in a very cosmic process to the compost pile, the earth, and plants. In the farm organism, dandelion stimulates the cosmic nutritional stream through the silica processes while the chamomile preparation, through the limestone process, stimulates the earthly nutritional stream.

506: Dandelion Compost Preparation

AP: *Throughout history, dandelion has been associated with both the liver and Jupiter. Would you elaborate on Steiner's insights regarding dandelion's relationship to Jupiter and the liver?*

HC: In the compost pile the dandelion compost preparation may help to cleanse or break down into basic elements some of the toxic substances that might be in the manure or plants, much the same way the liver and dandelion cleanse the human body. Beyond that breakdown, there can also be a building up into humus compounds. It is much the way that food in our body is totally destroyed and then built up into substances compatible with our body. Another reason for choosing the dandelion flower is its yellow color and elsewhere in the Agriculture lectures, Steiner talks about how the various outer planets influence the color of flowers in the plant kingdom. In particular, yellow and white flowers are related to Jupiter.

AP: *For someone new to biodynamics, would you briefly describe how you make the dandelion preparation BD506 at JPI?*

HC: One must first obtain the necessary materials: dandelion blossoms, and the peritoneum. The dandelions are harvested in the morning after the dew has evaporated and up to an hour before meridian noon (1:30 p.m. daylight savings time in our location). We usually start at about 9:30 a.m. and pick blossoms until as late as 12:30 p.m. Flower (air) days are chosen when possible, but we need so many blossoms at JPI that we don't have that luxury, especially when considering the weather. It is more important to pick only the blossoms with a firm center core, usually about one forth inch wide, resembling a "bull's eye." This is usually the first day that the blossom opens. If one picks fully opened blossoms with no center core, they are past their prime and less effective, and they will frequently turn to fluff in the drying process. If one has a peritoneum available, one would let them wilt for a few hours before stuffing them into the sheath. If you are stuffing the sheaths in the spring, one could hang them up in the summer sun much the same way one hangs up the deer bladders containing yarrow. I have done this in the past with excellent results as far as I can tell.

Since we do not slaughter until fall, we dry the blossoms and store them in a cool, dark, dry place until then. The dandelions should be dried quickly in a warm place with air circulation but not in the sun. We dry the blossoms on fiberglass screen drying racks in the attic. Excessive heat or prolonged drying will cause the blossoms to ripen to fluff. They should be checked daily. If it is necessary to store the blossoms for a lengthy period before actually inserting them in the peritoneum or mesentery tissue, our preference is to vacuum seal them to prevent damage from insect pests. Considering dandelion's relationship to Jupiter and knowing that tin is associated with Jupiter, the best surface on which to dry dandelion may prove to be made of tin to more fully utilize the Jupiter influence.

Top: Twenty-four pounds of dried dandelion were moistened and sewn into "pillows." Bottom: Hugh Courtney sits at the edge of the pit where the dandelion-stuffed "pillows" will be buried. The pit is approximately 18 inches deep.

I don't know where you would find pure tin screening or perforated tin. Tin, as we usually know it, is alloyed with other metals, and those would most likely be detrimental to the process. In anthroposophic medicine, remedies using dandelion will often have the dandelions grown in soil to which metallic tin has been deliberately added.

When we are ready to stuff the peritoneum or mesentery, we moisten the blossoms with a warm, not scalding, tea made from dandelion leaf. The blossoms should be thoroughly moistened but not so much that you can wring water out of them. We then make pouches for the blossoms

506: Dandelion Compost Preparation

by cutting the peritoneal tissue into rectangles, folding them in half, and always keeping the interior of the tissue on the inside of the pouch. We stitch up the two sides and part of the top. The blossoms are packed in tightly, creating an irregular pillow approximately six inches across and about three or four inches thick. This seems to be a manageable size, especially for the workshops. We bury the pillows in a pit that is no deeper than eighteen inches, in the fertile layer of the soil, not the subsoil. The pit is lined with hardware cloth to keep animals out and to prevent shovel damage when digging them up. A layer of fiberglass screening is placed between the hardware cloth and the pillows creating an envelope or package of sorts, which prevents the loss of valuable material when we retrieve them in the following year.

We also try to place a two-inch layer of peat moss between the hardware cloth and the top and bottom of the fiberglass screen envelope. About eight inches of earth is shoveled on top of the contents of the pit. The dandelion preparation needs to spend the winter in the earth to be enlivened by the winter process during that period when the Earth is most inwardly alive.

At JPI we keep the dandelion preparation in the ground until the following fall so that students at the fall workshop can experience both the burial and the retrieval. By making the preparation, they can observe the original materials and when we dig them up they can experience the transformation. In ordinary biodynamic practice, the time for unearthing the dandelion preparation would be in the spring, definitely after Easter, with our preference being around Ascension Day.

After the preparation is dug up, the sheath is removed and the transformed material is rubbed over a quarter inch screen and broken up into more consistent particles. Sometimes, it may be too wet to screen right away so we let it dry for a day or so. The preparations are then stored in a lidded, glazed clay crock surrounded by peat moss in a wooden box in the root cellar.

AP: In her book, The Biodynamic Year, *Maria Thun recommends using all of the BD plants in blossom and plant teas to enhance health, yield, and taste. There is no mention of using the compost preparations individually. Have you experimented with using the dandelion preparation BD506 by itself?*

HC: We have used it as a seed bath and as a spray to increase yield. These are seat-of-the-pants experiments and wouldn't qualify as scientific. One experiment was written up several years ago. We grew snow peas and sprayed half of the row with the dandelion preparation. We used one unit of BD506 with two quarts of water that we stirred/potentized for ten minutes in a one-gallon jug using a lemniscate motion. We sprayed two times on fruit days. The yield per plant was much greater from the dandelion-treated plants. This year, we did an experiment with green beans using seed baths on a fruit day. Dandelion was used as one of the seed baths. The seed baths were potentized for twenty minutes using a half-gallon apple juice jar and using a back and forth and side-to-side motion that creates a lemniscate or figure-eight movement of the water in the bottle. The water for the control was also potentized in the jar. I did four different treatments, with a control of water only, dandelion (BD506), oak bark (BD505), and a combination of the dandelion and oak bark preparations. Where the dandelion preparation was used as the seed bath, the yield was much greater than the control and modestly better than the oak bark alone. The combination of the dandelion and oak bark was slightly better than the dandelion alone. We took bean-counting to an extreme, counting both the pods and the beans at four stages of development, pulling twenty-five plants from each bed each time, and arriving at a total of one hundred plants from each bed. I haven't done an analysis of all the numbers yet but hopefully, it will be written up in the newsletter at some point. It doesn't qualify as sound science but it gives a reasonable picture of the influence of dandelion.

If we take the image of Jupiter's traditional association with wealth and abundance, and take that thought form and apply it to the question of yield, I think that we will find in the future that it will prove out quite nicely in increasing

506: Dandelion Compost Preparation

yield. Biodynamics has a reputation for great quality but modest yields when compared to conventional agriculture. I think that if we use the dandelion preparation more intensively, by itself (in conjunction with the usual application of all the other preparations, of course), it would go a long way to improving yields. That is an area for a lot more research.

In several places, Steiner states that the preparations give the plants the awareness and the means to draw from their surroundings the kind of nourishment that they need. It is that picture, of the preparations giving the plant kingdom the wherewithal to pull to itself what it needs for its nourishment, its growth, that led me to the sequential spray process. If the preparations really did work like this, and since we were experiencing a serious drought, I thought that if I chose the right preparation then I could help the plant to draw the much-needed moisture to itself. I didn't know which preparations to use so I used them all in a specific order. And lo and behold we got rain. It was very scary at first. The initial experience was a bit of a Sorcerer's Apprentice kind of experience. By studying the history of the biodynamic plants and the qualities associated with their ruling planets, we can get clues and perhaps intuitively perceive how to augment their use in a beneficial way. In biodynamics, we are barely touching the surface of what all the preparations can bring to agriculture.

AP: Is there anything else you would like to add?

HC: For a deeper understanding of the preparations and their contribution, I highly recommend the book *Earth and Man* by Karl König. König does one of the absolute best jobs around of trying to enlighten us to a better understanding of both the silica process and the limestone process. *Earth and Man* is one of my favorite biodynamic books. I think I'm reading it for not less than the third or fourth time. In *Sub Nature and Super Nature*, Pfeiffer gives a profound lecture on the cosmic and earthly nutritional streams. I recommend that book, as well.

AP: Thank you, Hugh.

507

Valerian officinalis
Valerian

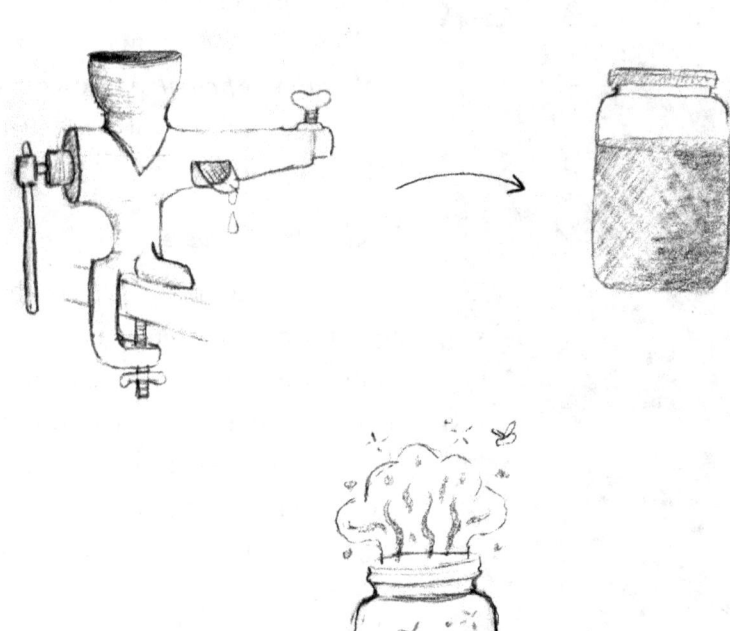

507: How to Make the Valerian Preparation

Patricia Smith and Hugh J. Courtney
Applied Biodynamics, *no. 29/30, 2000*

The following describes the steps we usually take when making the valerian preparation at the Josephine Porter Institute in Woolwine, Virginia. Valerian (*Valeriana officinalis*) grows two to five feet tall in moist soils with light shade to full sun.

Picking the Flowers

What to look for: White flowers with petals a pink hue to pure white. Refrain from picking those that are still a bit green and those that are past bloom that go a little brown on the edges. For harvesting, you can use scissors, a knife, or fingernails. When cutting, leave a little bit of stem.

Top: *Valerian (Valeriana officinalis) begins to form flower heads in the spring. Bottom: With long, sturdy green stems, valerian (Valeriana officinalis) grows two to five feet tall and prefers moist soils. (photos ©JPI staff)*

507: How to Make the Valerian Preparation

Time of Year to Harvest

May through June (Northern Hemisphere). If possible, cut it every nine days until it finishes blooming.

Time of the Moon

Preferably in a flower period, that is, when the moon is in one of the air/light constellations of Aquarius (Waterman), Gemini (Twins), or Libra (Scales). Ideally, the moon should be ascending (Aquarius or Gemini) or approaching full (Libra) as well.

Time of Day

If there's not too much dew on the flowers to contend with, it is fine to start at sunrise. Otherwise, picking the flowers from 8:30 to 11:30 am (daylight saving time) is a good guideline, not too wet, and yet not too dry. Picking should be completed not less than one hour before meridian noon.

Top: When picking valerian (Valeriana officinalis) flowers, look for petals with a pink hue to pure white. (photo © JPI staff)
Middle: Harvested valerian flowers to be used for making BD 507.
Bottom: Misting with a little water helps to ease the valerian blossoms through the grain grinder. (photos © Patricia Smith)

Processing the Flowers

Once flowers are picked, they are ready to go through a grain grinder or food processor. Place a bowl under the grinder to catch the processed valerian. Gently mist flowers with a little water to ease their way through the grinding process. Scrape the grinder consistently with a small knife rather than losing some of the paste when using one's fingers. When processed through, the valerian is the consistency of pesto, a lovely bright green paste.

Extracting the Liquid

When all flowers are ground, use a juice press or wheatgrass juicer to extract the liquid. Alternatively, a good mechanic could probably build a small hydraulic press with a few materials at far less cost.

The ground valerian paste is placed into a cotton cloth. Recommended are two handkerchiefs, enclosed and folded loosely over

Top: Use a small knife or similar tool to scrape, rather than losing some of the paste or flowers when using one's fingers. Bottom: Misting with a little water helps to ease the valerian blossoms through the grain grinder. (photos © Patricia Smith)

507: How to Make the Valerian Preparation

the valerian paste, and placed into the container of the press. The handkerchiefs should be pre-moistened and then wrung dry by hand. Loosely fitted works best, because if it is wrapped too tightly, the pressing force will tear the cloth apart, causing the valerian paste to squeeze out. The extracted juice is then collected in a sterilized bottle. To sterilize, glass bottles should be submerged for up to ten minutes in rapidly boiling water. "Grolsch" style brown beer bottles with an adjoining cap work well for this process. Use a sterilized funnel so you will not miss a drop; every little bit counts. One needs to ensure that all equipment is sterilized so that stray fungus or yeast spores will not contaminate the valerian juice. The press can sit for one to three hours as every drop is collected.

Top: The juice collector and wooden block are parts to the K & K Juice Press. The handkerchief is used to enclose the valerian paste. Middle: A wheatgrass juicer may not extract as much liquid as a more refined juicer. However, it does grind and extract juice at the same time, suitable for making a small batch of BD 507. (photos © Patricia Smith) Bottom: The paste of ground valerian is placed into a cotton cloth. (photo © JPI staff)

A simpler method for small-scale juice extraction is to place the moist paste in the handkerchiefs and twist them until the juice starts dripping through the bottom of the handkerchiefs into a bowl. Quantity extracted will be much less than with a hydraulic press, but may be suited for a small-scale operation since it takes only 20 drops or so to be able to treat 10 to 15 tons of compost.

Fermentation Time

When finished with the extraction process, remove the funnel and seal the opening of the bottle with a sterilized fermentation lock (if plastic, scald for one minute in near-boiling water). Label each bottle with the date, moon aspects, moon phase, and what type of press is used. After six weeks, the fermentation lock can be removed and a regular cap can seal the bottle. Fermentation for valerian

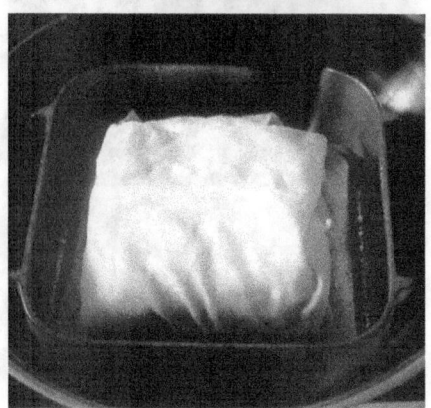

Top: After the valerian flowers have been ground, enclose the paste into a double layered handkerchief or similar cotton cloth.
Bottom: Fold cloth loosely; if it is wrapped too tightly, the pressing force will tear the cloth apart, causing the paste to squeeze out.
(photos © Patricia Smith)

507: How to Make the Valerian Preparation

is six weeks to three months, so use your judgment for sufficient time.

The Filtering Process

From this point on you can decide whether to filter out the sediment, or use your fermented valerian preparation BD507 as it is. For filtering the sediment, we use two glass manual drip coffee makers, milk filters, and Melita coffee filters. Be sure to sterilize coffee makers and plastic filter holders in boiling water. Pour the fermented valerian through the first coffee maker, lining the filter holder with the milk filter. Once the filter starts to clog, start the second coffee maker, also lined with the milk filter. Once all the fermented valerian has dripped

Top: The K & K Juice Press extracts valerian juice with little sediment compared to a wheatgrass juicer. Use a funnel to catch every drop. Bottom: After extraction in the K & K Juice Press, the paste removed from the cloth resembles a flat green "cake." This cake can be crumbled and re-moistened with a little water to return it to a paste-like consistency. The re-moistened paste goes through the juice press again and is funneled into a separate bottle for a valerian water extract. This can be used for an emergency backup if the pure valerian extract supply is depleted.
(photos © Patricia Smith)

through, remove the sediment-laden filters and replace them with Melita-style coffee filters, a less porous filter than the first. Once the fermented valerian has dripped through, most of the sediment is now removed. Remove filters and holders from the coffee maker. Using a sterilized funnel, pour the contents of the filtered fermented valerian into sterilized glass bottles. These bottles can now be stored for years in a cool dark place. From time to time (every three to six months), uncap the valerian to release fermented gases. This is important, because gas buildup in bottles of fermented valerian may explode and shatter when left unvented.

For Use

One unit of fermented valerian is one to two milliliters. To convert milliliters to teaspoons, use the formula: 5 milliliters x .2 = 1 teaspoon. One unit of fermented valerian is usually added to about two gallons of water. Once stirred into water for the recommended ten to fifteen minutes, or even up to twenty minutes, uses include distributing to a compost pile, as well as seed soaks and frost protection, to name a few. For further information on its uses and applications, see the next article, "The Valerian Preparation: Additional Perspectives."

See https://jpibiodynamics.org/pages/resources.

The Valerian Preparation:
Additional Perspectives

Hugh J. Courtney
Applied Biodynamics, No. 29/30, 2000

In our effort to cover as completely as we can everything you ever wanted to know about BD507, the valerian preparation, it would be useful to view historical and other perspectives on the making and use of this preparation. In the first place, no other preparation receives such little attention in Rudolf Steiner's description of the various preparations as does valerian. It is almost as an afterthought that he describes how one makes and uses this preparation, and it is a good possibility that too many easy assumptions have been made about Steiner's indications for the valerian preparation. So far as I can determine, Steiner had little or no opportunity to clarify any of the questions that may have arisen in the minds of those early biodynamic pioneers concerning either the making or the actual use of the valerian preparation.

It is my understanding that three people were principally involved with the initial making of the various preparations: Ehrenfried Pfeiffer, Günther Wachsmuth, and Immanuel Voegele. Much of this initial effort took place before Steiner gave his lectures, but Steiner kept the instructions to the bare essentials as far as these individuals were concerned. He also did not give a great deal more elaboration when he presented the lectures in Koberwitz, most especially concerning the valerian preparation.[1] Given the fact that Steiner died in 1925, less than a year after the agriculture course was given, and was quite ill

---
1   See Keyserlingk, *Developing Biodynamic Agriculture*, p. 43.

for the last six months of his life, it is probable there was little chance to clarify his indications about the preparations.

Within a few short years after the agriculture lectures, it appears that the general practice for making the preparation essentially involved the practice described elsewhere of gathering blossoms, grinding them up, and then extracting the juice for subsequent fermentation. However, at least a few practitioners also chose to prepare valerian in a fashion similar to the making of a flower essence. That is, relatively few blossoms were placed in water in a bottle and then hung up outdoors to receive the benefits of sunshine for a specific time. I should point out that the former method used at JPI is one received from Josephine Porter, who received it directly from Ehrenfried Pfeiffer, who presumably received it directly from Rudolf Steiner.

Concerning the use of the valerian preparation in its role as a compost preparation, the general practice in Europe seems to be that it is only sprinkled on the top of the compost pile after the other preparations have been inserted. It seems that it is always applied at the time the pile is treated with the other preparations. The Central European technique is thoroughly described in the following extract from the book, *Bio-Dynamic Farming Practice*:

> Based on many years of experience, the German Research Group for Biodynamic Farming Methods (*Forschungsring für Biologisch-Dynamische Wirtschaftsweise*) has published the following brief instructions for using and storing the compost preparations:
>
> The six preparations are made from the medicinal plants yarrow, wild chamomile, stinging nettle, oak bark, dandelion, and valerian. They are added to compost, manure, dung liquor and slurries. To treat a compost heap, stacked manure or deep litter, use a stick to make five holes, each about 50 cm deep. The distance between holes should be not less than 50 cm nor more than 2 meters. Put 2 or 3 grams (a small teaspoonful) of the five solid preparations singly into consecutive holes, as shown in figure 32. The sequence does not matter that much, it is used for

The Valerian Preparation: Additional Perspectives

practical reasons. The preparations are quite light in weight; to make sure that they go down to the bottom of their respective holes, mix each with a little dung or moist earth, shape into a small ball and drop down this hole. Close the holes up, so that the preparations are completely surrounded by manure. For the liquid valerian preparation, you'll need hand-hot water, which should be soft or have been left to stand for some time. Use 2 to 3 ml (a small teaspoonful) to 5 liters of water and stir for 20 minutes, reversing direction at intervals (clockwise until a vortex has formed, then counterclockwise, etc.). Spray the surface of the heap with the solution.[2]

As biodynamic agriculture moved farther to the west, there was added the practice of inserting the valerian in its own hole in the pile along with a sprinkling on the exterior surface of the pile as detailed by Katherine Castelliz in *Life to the Land* as follows:

> These six plants are used for the making of the six compost preparations BD502–BD507. The study of these plants may help to a better understanding of the working of these preparations. As compost preparations they are used in order to transform garden refuse as well as animal manures into substances that advance the development and health of plants. The fact that organic substances are transformed into good compost much quicker under the influence of the preparations than without them shows that the substances are really "prepared" as "plant foods" and losses are kept to a minimum.
>
> The heaps, either manure heaps or heaps made of vegetable matter, should measure not more than four yards (or meters) across but can be much narrower. Their height should not exceed five feet (1.30 meters). Their length is unlimited. The mass of vegetable matter or manure will rot better and quicker if some soil, or better still, some half rotted or rotted compost is incorporated in the heap. Vegetable matter has to be set up with dustings of hydrated lime (or ground limestone) between layers. The sides of the heap should be slanting, which makes it wider at the bottom than it is at the top. The preparations are put in at a distance of about one yard (or 1 meter) along

---

2   Sattler and Wistinghausen, *Bio-Dynamic Farming Practice*, pp. 87–88.

the upper edge of the heap. One makes holes with a stick or a crowbar, about 20 inches (50 cm) deep, all the way around the heap. Small heaps get three holes in each side. One portion (1 gram) of preparation is put into each hole, and each hole receives a different preparation. Start with BD502 and go on to BD507, then begin again BD502 with until all holes are served. Never put all the preparations into one hole. When all is done fill in the holes to close them. To make (valerian) ready for use put enough of the liquid into lukewarm rain water to color it slightly, make sure you can smell the valerian in it, and use some of it to pour into the appropriate hole or holes and use some of it to sprinkle over the heap after all the holes are closed. Cover the heap with a thin layer of soil or with peat, lawn clippings, straw, etc. Manure heaps should have a generous covering of straw. Do not set up manure from different animals in different heaps, but mix them, unless you need some special manure for a particular job. For 9 to 12 cubic yards (or meters) of a finished heap one set of preparations is required.[3]

Interesting enough, Katherine Castelliz concludes the above description with a footnote saying, "See also George Corrin, *Handbook on Composting and the Bio-Dynamic Preparations*." In that book, Corrin describes using valerian applied essentially only to the exterior of the compost pile.

The practice of putting the valerian within as well as on the pile apparently originated with Ehrenfried Pfeiffer. It should be noted that Katherine Castelliz worked very closely with Pfeiffer in Dornach for a considerable time. The present instructions for using the valerian preparation remain just as they were received from Josephine Porter in 1984, and were given to her by Ehrenfried Pfeiffer himself. There is no record available to suggest that Pfeiffer ever discussed the placement or use of this preparation with Steiner, but one must presume that he would have had that opportunity. Since Pfeiffer seems to have had at least some ability for direct communication with the elemental beings, one should not dismiss as a total

---

3   Castelliz, *Life to the Land*, pp. 27–28.

The Valerian Preparation: Additional Perspectives

aberration his use of valerian as a preparation to be placed within the pile as well as on it.

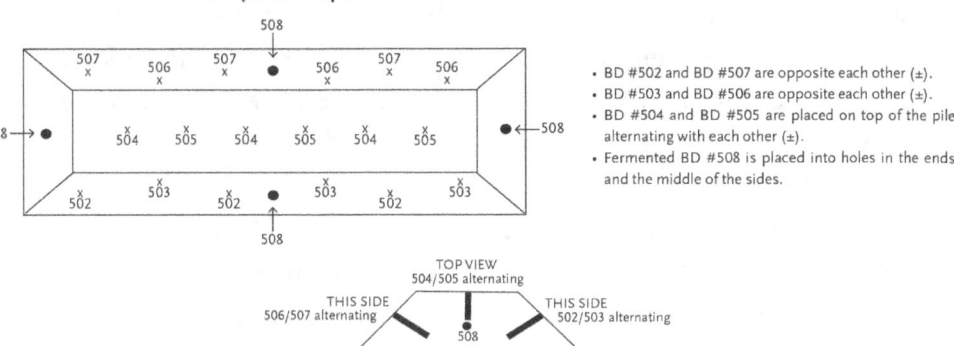

Given Rudolf Steiner's clear indication that there are differences in the geographic forces in North America insofar as how tightly the manure should be packed into the cow horn,[4] it is possible that Pfeiffer also discerned the need to use valerian differently as he moved farther west. Steiner gave lectures on the differences in the manifestation of forces in different areas of the Earth, published as *Geographic Medicine*.[5] I believe that every biodynamic practitioner who wishes to broaden an understanding of the forces within the various biodynamic preparations should study these lectures and how they could impact the farm.

Those whose biodynamic training originates in Europe, but whose practice is accomplished in America seem to still follow the technique of treating the exterior of the pile only. Those whose biodynamic training originates in America, where Pfeiffer's influence was paramount, follow the practice of both exterior and interior application of BD507.

---

4 Steiner, *Agriculture*, p. 79.
5 Currently published as the first two lectures in Steiner, *Secret Brotherhoods and the Mystery of the Human Double*.

It is also worth bringing to the reader's attention the view that Harvey Lisle held with respect to the valerian preparation. He felt quite strongly that BD507 should be placed only within the pile, as he believed that sprinkling the pile with the valerian effectively "kills" the biodynamic forces in the pile. Lisle was a biodynamic practitioner for many years and served for several years as the secretary for the Biodynamic Farming and Gardening Association board of directors.

Whereas many would be inclined to discount Harvey's findings because he used his dowsing abilities to arrive at his conclusions, I have too much respect for his years of biodynamic efforts not to at least pause and seriously contemplate any thoughts he may offer out of his experience. I am inclined to believe that Harvey may not be asking exactly the right question when he concludes that valerian sprinkled on the exterior of the pile "kills" the biodynamic effect within the pile. Nevertheless, this is another of those questions in biodynamic agriculture that cries out for comprehensive research. As I already observed, the farther west that the biodynamic movement travels, the more the view seems to call for placing this preparation within the pile, and perhaps eliminating it as a surface application entirely.

When one tries to understand the underlying reasons as to why Pfeiffer would have chosen to also insert the valerian within the pile, we are provided a clue from reading Karl König. In *Earth and Man* he states: "Although a tremendous amount of lime is precipitated in our skeleton, the form into which it is precipitated is silica."[6] Later, in answer to a question, König says further:

> When you study the silica process in man you find that it is always on the surface. It is not only on the outer surface, but also on the inner surface. Silica is distributed in the hair, and all over the body in the layers of the skin, where horny material is continually forming and being discarded. It is a layer in the eyes

---

6 König, *Earth and Man*, p. 240.

and in all the openings that form our senses. Now consider the skeleton. The inner and outer surfaces of the skeleton are also filled with very fine silica substance.[7]

It is quite possible that Pfeiffer, when faced with the very strong earthly forces in America, felt that the cosmic forces needed to be within the compost as strongly as possible. Again, this is a matter calling for research. When we take a look at uses for BD507 other than as a compost preparation, it appears to receive more attention as a preparation for individual application than any of the other five compost preparations. An additional excerpt from *Bio-Dynamic Farming Practice* describes some of the other uses of valerian as follows:

> A cold extract made from the flowers of valerian (*Valeriana officinalis*, BD507) gives compost and manure the power to relate properly to phosphorus. Added to a manure or compost heap, the valerian preparation not only regulates phosphorus and temperature processes but also surrounds the heap with protective warmth. This quality can be put to good use for the prevention of late and early frost damage. If there is a serious risk of night frosts, spray the crops—e.g., flowering soft fruit plants or sensitive species such as beans, tomatoes, early potatoes and also basil, late in the evening with a valerian solution. As a rule, there will be no damage with temperatures dropping to $-3°$ or $-4°C$; if they go down lower than that, damage cannot be prevented but can at least be minimized to some extent. If plants are covered with hoarfrost in the morning, a second application of valerian will often save the situation. If valerian is sprayed on flowering plants (e.g. pulses, rape, linseed and mustard) and fodder plants grown for seed (e.g. sainfoin, phacelia and many others) when the shoots are c. 15–20 cm high and again just before they come into flower, the flowering process will be greatly stimulated. Crops will flower abundantly and at a good pace in sunny weather, ensuring a greatly enhanced yield.[8]

---

7   Ibid., 246.
8   Sattler and Wistinghausen, op. cit., pp. 86–87.

Beyond its uses for frost protection and to stimulate flowering, the valerian preparation is further recommended by Castelliz as follows:

> Plants in special need of light and warmth (tomatoes, beans, carrots, etc. should be sprayed with this valerian extract in cool and wet summers. The general health and resistance to disease are being promoted in all plants that are treated with BD507, the color of flowers is deepened, the aroma of fruit is enhanced. Use one portion (1 milliliter) to a bucket of water.[9]

Perhaps the greatest benefit to be gained from independent use of BD507 is as a seed bath when starting seeds to improve germination and overall vitality. Among the several preparations used for seed baths, valerian is perhaps the most versatile. It is recommended for a long list of crops beginning with wheat, corn (maize), beets (including fodder and sugar types), onions, leeks, celery, carrots, celeriac, tomatoes, peppers, potatoes, cantaloupe, pumpkin, and other fruiting crops.[10]

Lastly, when BD507 is used either alternatively or in conjunction with BD501, fruit set and yield can be substantially enhanced. At least one vineyard experienced a full point improvement in Brix readings for their grapes this fall when using a combination of BD501 and BD507. Given the cool, wet summer and early fall experienced at the vineyard, this was very helpful in being able to harvest the grapes in a timely manner. Even more important than using valerian by resorting to a checklist approach, as in the case of seed baths, for instance, is to understand valerian as a carrier of forces rather than simply as a substance. One of the most helpful pictures in developing such understanding is to be found in C. B. J. Lievegoed's "Working of the Planets and the Life Processes in Man and Earth," in which he makes the following statements about valerian:

---

9   Castelliz, op. cit, p. 32.
10 Hugh Courtney, "Seed Soaks with the Biodynamic Preparations," *Applied Biodynamics*, no. 7, spring 1994.

The Valerian Preparation: Additional Perspectives

Rudolf Steiner says about this preparation little more than that it stimulates the plants to behave properly toward what is called phosphorus. Valerian is not prepared in any special way. The flowers are pressed and the extraction highly diluted with water. The mixture is sprinkled over the compost heap. There is no need here to prepare the plant within the sphere of the Earth. Diluting and potentizing are enough, because we are dealing with the most distant forces of Saturn, so close to the spirit. These must be communicated to the heap, and by sprinkling them over it we surround it with something like a spiritual sheath, just as Saturn encloses the whole planetary space.

How can we understand this "to behave in the right way in relation to what we call the 'phosphoric' substance?" Rudolf Steiner describes the medicinal effect of phosphorus as strengthening the "I" (in the case of the plant this means the spiritual archetype), as against a too-strong astral activity. With the oak bark preparation, we achieved that the astral should work as a regulator against the rampant ethereal. Here, in the case of valerian, it is a spiritual element that must win the upper hand, so that the plant may, again and again, realize its own archetype. We can test the medicinal effect of valerian by taking a large dose (e.g. 50–60 drops) of *Tinctura valerianae*, made from the root. We notice a slowing down of the pulse, a leaden heaviness in the limbs, and a lowering of all functional rhythms. This remedy is found in every medicine chest and is used in all cases of nervous excitement and insomnia. With the valerian, we have enumerated the last of the six compost preparations and have brought its effect upon plants into relation to Saturn, most distant from the Earth and closest to the spirit.[11]

At this point, I would like to take a small exception to the statement of Lievegoed that "valerian is not prepared in any special way." After the extraction of the juice from the flowers, the juice goes through a fermentation process before use, which can be construed as a "special way" of preparing it. The alternative "flower essence" method for making this preparation can also be construed as a "special way" of

---

11 Lievegoed, "The Working of the Planets and the Life Processes in Man and Earth," p. 31.

preparing it. Both fermentation and the "flower essence" treatment are means to imbue the preparation with a greater affinity for a spiritual rather than an earthly dimension.

Another useful picture in helping us gain an understanding of the force level rather than the substance level at which valerian functions is to be found in reading the following from *Earth and Man* by Karl König:

> We now cover [the compost heap] and sprinkle it with valerian. Why do we do this? We enliven the phosphorus process, and this is a process that calls down the help of the heavens to aid all that has been done. To call down the forces to act in the way we had intended these forces to act—this is actually done with valerian.
>
> I could of course now speak at length about valerian as a medicine for men and animals. But here you use it not so much in a medicinal way, but much more generally. It is done—I hardly dare to say it for fear of creating a wrong impression—as one uses incense at the altar. You do something to call down the help of the higher worlds: "Let the archetypal plant come down."... The sprinkling of the compost heap is actually the most important process, and should live in our souls when we perform it. Something like the sprinkling of valerian, this calling down of the help of the spirit for the work on earth, should penetrate and infiltrate our whole existence as gardeners and farmers....
>
> We must realize more and more what a holy gift we carry as we learn to practice it (biodynamic agriculture) and to understand it.[12]

No matter how we choose to make the valerian preparation, or how we apply it on or within a compost pile, it seems to me that the attitude of reverence for a sacred act, which König suggests, is the most important approach to using valerian within the Earth.

---

12   König, *Earth and Man*, pp. 310, 312.

508

Equisetum arvense
Comon Horsetail

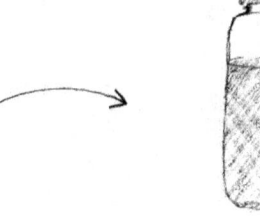

508: Revealing the Hidden Forces of Equisetum arvense

Hugh J. Courtney

Applied Biodynamics, *spring 2010, no. 68*

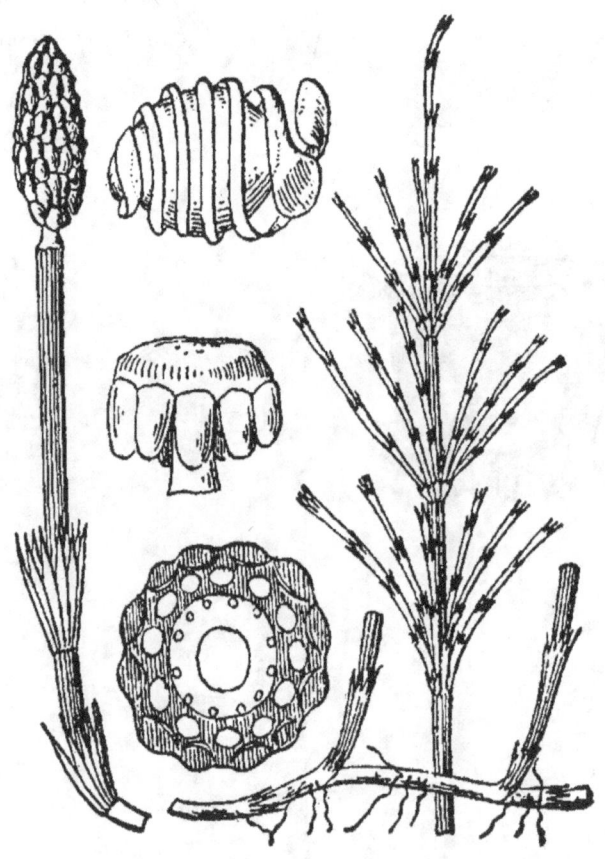

Equisetum arvense

508: Revealing the Hidden Forces of Equisetum arvense

When I began distributing the preparations following Josephine Porter's death in 1984, I was troubled by the fact that two of the preparations, as I understood them from my training by Josephine, seemed to be of virtually no interest to anyone ordering preparations. Those two were BD501 (Horn Silica) and BD508 (*Equisetum arvense*, or Horsetail Herb). The problem concerning BD501 turned out to be that most growers simply could not work it into their busy springtime schedule and, since authorities from both Europe and Australia presented cautions about the strong light/sunforces in North America and the danger of plants being easily burned after applying BD501, growers were given an excuse not to use it.

Much of our educational efforts at the Josephine Porter Institute for Applied Bio-Dynamics has been devoted to overcoming the reluctance to use BD501, since we strongly believe that this preparation needs to be used to the maximum extent possible. Provided one uses proper care to avoid application too long after sunrise, the use of BD501 should not bring about the negative effects assumed by authorities from other parts of the world. However, it turned out that there was a much greater reluctance toward applying BD508, *Equisetum arvense* (here referred to as BD508, or simply Equisetum) than I initially had any reason to believe. At one point, I was taken to task by a member of the board of directors of the Biodynamic Farming and Gardening Association for having the temerity to refer to Equisetum as a "biodynamic preparation" and labeling it as BD508. This individual, trained in biodynamics in Europe, shared the view that Equisetum was appropriate only when one was faced with a problem of fungus. In such instances, Equisetum was brewed into the fresh tea and applied as a prophylactic to curtail or eliminate the fungus problem.

My understanding of Equisetum was influenced by Lily Kolisko's description in *Agriculture of Tomorrow* of a fermented tea recipe, and in particular by the Capillary Dynamolysis (Rising Picture), which contrasted the fresh tea and the fermented tea recipes. It was abundantly evident to me that the fermented tea recipe was a much stronger carrier of forces than the fresh tea recipe. Moreover, such fermented tea goes through at least the same degree of preparation as does the BD507, or Valerian preparation.[1] We adopted the view at JPI that BD508 yields much better results in the fermented-tea version, and that it should be applied especially as a soil spray early in the growing season (shortly after a spring application of BC and/or BD500). Thus, Equisetum can be preventative before the fact of fungus, rather than merely being used as a treatment when fungus is already present or its appearance is almost certain. The antipathy toward the use of Equisetum prompted me to begin looking more deeply into what Steiner had to say about it. Initially, I looked at the organization of the agriculture lectures themselves. The first three lectures provide the basis for what we call the biodynamic agricultural preparations.

In his *Agriculture Course*, lecture 3, Steiner promises to provide us with the means to administer a "dose of clay" to bring about a mediating force between the silica and limestone polarities. Although numerous individuals have experimented with this or that "new" preparation to bring about this seemingly unfulfilled promise of Steiner's, ranging from Hugo Erbe's clay preparation as well as various versions of a "horn clay," I would suggest that it is not a specific clay preparation to which Steiner was referring, but rather the nine biodynamic preparations that he did give to us. My understanding is that when Steiner described the "mediating force" (clay) that was needed, he met that need when he described how to make and use the nine biodynamic preparations. When we examine the structure of the course of lectures, Steiner uses lecture 4 to describe for us the making of the preparations of Horn Manure (BD500) and Horn Silica (BD501), as well as the reason for their being. In lecture 5 he lays the foundation and then tells us about the compost preparations

---

1   For more on Lily Kolisko's Capillary Dynamolysis research and findings, see https://www.alchemywebsite.com/kolisko.html.

508: Revealing the Hidden Forces of Equisetum arvense

of Yarrow, Chamomile, Stinging Nettle, Oak Bark, Dandelion, and Valerian (BD502–507). In lecture 6, he presents the final preparation in his description of making *Equisetum arvense* into a tea. Steiner devoted more time and energy laying the foundation for the Equisetum preparation than he spent on any other preparation, so it is clear that he assigned some importance to it by including it in the agriculture course.

The Course Structure

It may also be of interest to examine the structure of lecture 6 in the light of the indications Steiner gave on how one should present a lecture on Spiritual Scientific subjects. In at least two different lecture cycles, Steiner described how the lecturer should approach his or her presentation. One such cycle is published as *The Art of Lecturing*.[2] In this and a second cycle of lectures that I read many years ago at Josephine Porter's, the steps Steiner outlined for the presentation of Spiritual Scientific subjects were as follows:

1. Indicate and identify the nature of the problem.
2. Give background and build a foundation to fully elaborate the problem.
3. Provide a solution to the problem.

When we look closely at lecture 6 in this light, the problem can be described as how to deal with "harmful plants and animals and what are commonly called plant diseases" using insights gained into plant growth and the relating forces. Steiner then describes the role played by cosmic forces or influences in the growth of plants as well as the relationship of those forces to the animal kingdom. While identifying various cosmic factors involving both the several planetary members of the solar system as well as the individual constellations of the zodiac, Steiner repeatedly identifies the Moon as the

---

2   Steiner, *The Art of Lecturing*.

overwhelmingly primary celestial factor affecting plant as well as animal growth. This is illustrated by the following selected sentences from lecture 6:

1. "Here we have all the influences that come in from the Cosmos—from Venus, Mercury and Moon—and ray back again, working upward from below. Everything that works in the earth in this way causes the plants to bring forth what grows in a single year and culminates in seed-formation. From the seed a new plant arises, and a third, and so on. Once more then: everything that works from the Cosmos in this way shows out into the reproductive forces—into the sequence of generations." (*Agriculture*, p. 108)
2. "Now a large number of plants—notably those that we ordinarily count as weeds—are greatly influenced by the workings of the Moon." (p. 109)
3. "What do we know of the Moon in ordinary life? We know that it receives the rays of the Sun upon its surface and throws them back again onto the earth. We see the rays of the Sun reflected—we catch them with our eyes—and the Earth, too, of course, receives these rays from the Moon. It is the rays of the Sun that are thus reflected, but of course the Moon permeates them with its own forces. They come to the Earth as lunar forces, and so they have done ever since the Moon separated from the Earth." (p. 109)
4. "With the Moon's rays the whole reflected Cosmos comes on to the Earth. All influences that pour on to the Moon are rayed back again. Thus, the whole starry Heavens—though we may not be able to prove it by the customary physical methods of to-day—are in a sense rayed back on to the Earth by the Moon. It is indeed a strong and powerfully organizing cosmic force that the Moon rays down into the plant, so that the seeding process of the plant may also be assisted; so that

the force of growth may be enhanced into the force of reproduction." (pp. 109–110)

Spray BD508 — For Use on Foliage

Standard Instructions (E. E. Pfeiffer)

Keep dry and in darkness until needed. Part portions may be cooked, keeping same proportions. Fully effective only when used following other biodynamic preparations.

TIME
- During wet, damp, or sunless warm weather as preventative of or care for fungus diseases
- Only during warm time of year
- When there is no wind
- Not when a rain will soon wash it off

USE
- One portion of BD508 to one quart water (for cooking)
- Two gallons water for later dilution
- Same stirring equipment as for BD500

PREPARE
- By cooking slowly in covered vessel for twenty minutes
- Strain and dilute
- Stir as for BD500, but for only fifteen minutes

SPRAY
- With clean sprayer, as for BD501
- Using fine mist spray
- Within a few days after cooked and diluted
- Gradually more dilute the more frequently it is used

The preceding excerpts are only a small sample of the many statements that Steiner makes concerning the Moon and the various ways that its forces manifest on the Earth. Steiner is laying the groundwork here in the middle part of lecture 6, regarding how Moon forces affect pests and weeds. The solution provided is the use of "peppers" of the specific pest. However, I would contend that the various "pest peppers" will yield the desired results much more readily if they are fully supported by prior use of the biodynamic preparations, and most especially by the use of BD508.

You will find it to be a most interesting exercise, as I have, to read the entirety of Lecture Six and compile a list of all of Steiner's comments regarding the Moon in order to gain a complete picture of the role of the Moon and Cosmic forces. Indeed, an even better exercise would be to read the entire course itself with a focus on the Moon and, in essence, compile a résumé of how the Moon exercises its influences on the Earth and all the Kingdoms of Nature, as described through Steiner's unique perspective.

Near the end of lecture 6, Steiner presents us with the solution to the original problem posed—that is, how to deal with "harmful plants and animals and what are commonly called plant diseases" when he provides background of how to handle extreme Moon forces, especially when such forces are too strong. Not directly examined by Steiner in this lecture is the other side of the coin, how one is to deal with Moon forces that are too weak. The solution Steiner presents to the problem of Moon forces is to bring about balancing or harnessing of these forces using a counterforce, which he gives in the following two paragraphs:

1. "Thus, we see the forming of mildew, blight, rust, and similar diseases. The over-intense Moon-influence prevents what should work upward from the earth from reaching the necessary level. The true force of fertility depends upon the Moon's influence being normal. It must not be too intense. It may seem strange, but it is so: this result is brought about, not by a weakening but by an over-intensity of the Moon-forces. If we merely theorized about it instead of looking at the process, we

508: Revealing the Hidden Forces of Equisetum arvense

might reach the opposite conclusion, but we should be wrong. Perception shows it as I have now described it. What, then, should we do?" (p. 118)

2. "We must somehow relieve the earth of the excessive Moon-force that is in it. And we can do so. We need only perceive what works in the earth so as to deprive the water of its mediating power; so as to lend the earth more "earthiness" and prevent it from absorbing the excessive Moon-influences through the water it contains. We can achieve this result. Outwardly, it all remains just as it is. But we now prepare a kind of tea or decoction—a pretty concentrated decoction of *Equisetum arvense*. This we dilute, and sprinkle it as liquid manure over the fields, wherever we need it—wherever we want to combat rust or similar plant-diseases. Here again, very small quantities are sufficient—a homeopathic dose is quite enough." (p. 118)

Thus, a careful analysis of the structure of lecture 6 provides us with the possibility of seeing this Equisetum preparation in a wider sense rather than in the narrower more "orthodox" view prevalent in biodynamic agriculture.

Using Equisetum: Two Approaches

The customary introduction to the use of Equisetum is well illustrated by the basic instructions provided by JPI in the one-page leaflet entitled "Brief Directions for the Use of the Biodynamic Sprays," and also the booklet *Bio-Dynamic Sprays* by Herbert Koepf.[3]

The one-page leaflet was inherited by JPI from Josephine Porter in 1984 and the instructions therein were probably originated by Dr. Ehrenfried Pfeiffer shortly after he brought biodynamic agriculture to this country, and may, indeed, have been based on the standard practice in European biodynamic circles. The booklet by Dr.

---

[3] Koepf, *Bio-Dynamic Sprays*. Currently contained in *Koepf's Practical Biodynamics: Soil, Compost, Sprays, and Food Quality.*

Koepf was based on a pamphlet by Evelyn Speiden Gregg, which was likely compiled under Dr. Pfeiffer's direction in the earliest days of biodynamic agriculture in this country. Thus far, we have not been presumptuous enough to supersede those instructions with our own version, although we have attempted to solicit input from various practitioners with a view to revising instructions for all the spray preparations.

BD508 Fermented Recipe

The following is a slightly modified recipe of the one developed by Lili Kolisko. Take one unit of shredded Equisetum arvense, about 8 ounces by volume and 1 ½ ounces by weight. Boil in one gallon of water and simmer for an hour. Let cool and transfer it to a crock or other storage container with a loose-fitting lid. Store in a cool place (for example, a basement or cellar) and allow to ferment ten to fourteen days, until the "characteristic smell" develops as described by Kilisko. Strain out the remaining particles, pour the tea into a glass jug, and store it in a cool, dark place until needed. It can be stored for six or more months without losing effectiveness.

To use, add a half gallon of fermented tea to 4 ½ gallons of water. Stir for twenty minutes in the biodynamic fashion and use to treat up to 1 ½ acres. One unit of horsetail can thus treat up to three acres with a powerful effect equal to that achieved by the fresh-tea recipe on a single acre. The ratio for dilution is 1 part fermented tea to 9 parts water. The fermented BD508 should most likely be applied as a soil spray, whereas the fresh tea version is probably the preferred form for use as a foliar spray.

As will be seen from the instructions for BD508, it is identified as a foliar spray. Dr. Koepf specifies that "Horsetail is used against such fungal diseases as mildew, rust, monilia, scab, soil born pathogenic fungi." He recommends that "One sprays this tea frequently, especially on garden crops." Here the implication would seem to be that the Equisetum is regarded also as a foliar spray. When one examines the book *Bio-Dynamic Farming Practice* by Sattler and Wistinghausen,[4] which is used as the basic text in a four-year training program for beginning biodynamic farmers in Europe, it is interesting to note that absolutely no reference to identifying it as a biodynamic preparation is made. It is pointedly not referred to as BD508, and it appears to be grouped along with several other useful plant-based sprays, such as seaweed, stinging nettle, wormwood, and tansy. Immediate use of the tea as a preventative is recommended, although the statement is made that "...stored in a barrel. It will ferment in due course and turn into the liquor, which does not affect its usefulness." This source also indicates: "Do not add horsetail when stirring preparations BD500 and BD501" (p. 79).

In 1997, some eight years after the initial publication of Bio-Dynamic Farming Practice, another booklet, *The Biodynamic Spray and Compost Preparations: Directions for Use*, compiled by five European biodynamic practitioners, including Wistinghausen, provides a brief mention for Equisetum under the heading "Horsetail (BD508) Equisetum arvense, a biodynamic plant care Preparation," of which the following statements are worth noting:

1. "The high silica content in this plant can be released either by decoction or by fermentation and be used to regulate fungal growth."
2. "In autumn and spring this Horsetail tea may be sprayed on the soil to control potential fungal infection and, where crops are at risk, directly on the plants as a preventive measure."

---

4 Sattler and Wistinghausen, *Bio-Dynamic Farming Practice*.

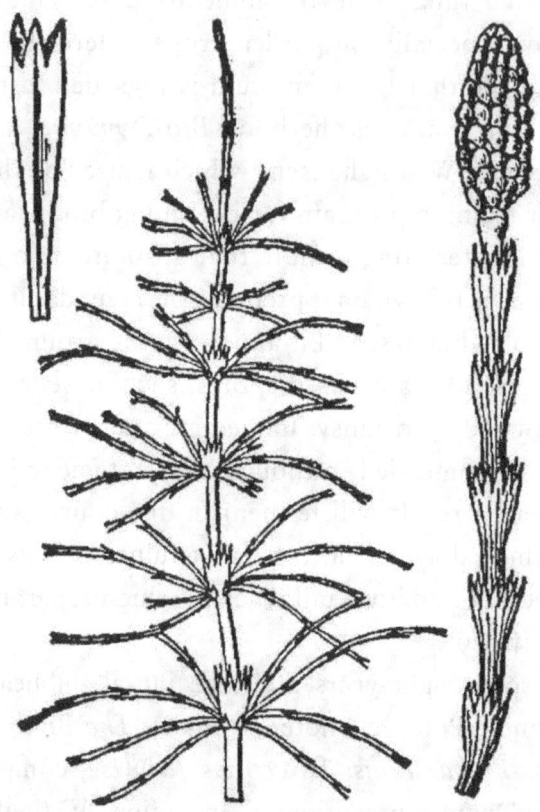

Equisetum pratense *(meadow horsetail)—Found in the northern U.S., Alaska and Canada. Moist woods, thickets and meadows. Grows up to twenty inches high.*

Thus, it is only after some fifty to sixty years or more of biodynamic practice that Equisetum is mentioned for application to the soil rather than just to plants, and its preparation by fermentation is endorsed. This delay occurred despite Steiner's very clear statement: "Nevertheless, you have a guiding line, and you will now investigate how equisetum works when you transform it as described, into a kind of liquid manure, and sprinkle it over the fields."[5] It is almost trans-

---

5   Steiner, *Agriculture*, p. 118.

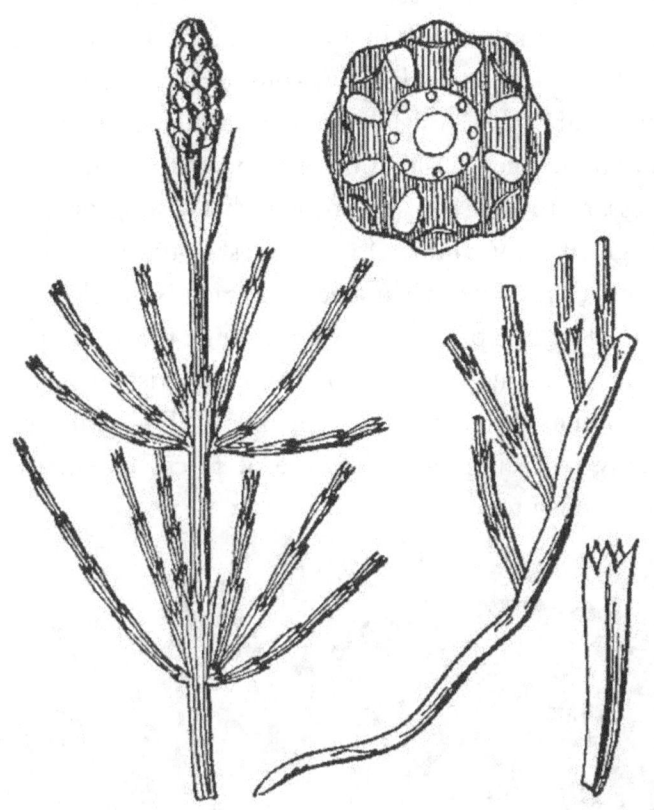

Equisetum palustre *(marsh horsetail)*—Found in North America, throughout California, Alaska and Canada. Nutrient-rich wet meadows. Grows up to two feet high.

parently evident that "liquid manure" is best arrived at by fermenting the Equisetum; and that it is the application to the soil that is meant when we are directed to "sprinkle it over the fields" as one would do if making any kind of manure application.

Although we have had the lectures Steiner has given us for almost eighty-six years, we sometimes do not read them carefully enough. The question to ask is, "Are we choosing to make our assumptions rather than paying attention to what he says?"

We at the Josephine Porter Institute have long relied on what we have gleaned from the book by Lily and Eugen Kolisko, *Agriculture of Tomorrow*. In that work, Lily Kolisko suggests that, when the Moon forces are too strong, we can "re-establish the right balance with equisetum." In describing the making of the tea, she states that "It is necessary to boil equisetum for rather a long time. The tea is of a light green color, and we found it best not to use it immediately, but in a few days' time, when a certain smell is developing." This is accompanied by two pictures derived through the technique of Capillary Dynamolysis, matured until the "characteristic smell" develops. It was clear that the matured ("fermented") Equisetum provided a dramatically more impressive picture of forces or energy at work than did the fresh tea.

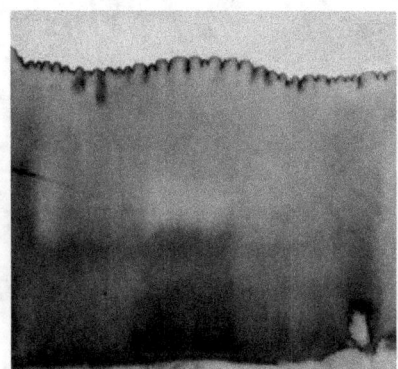 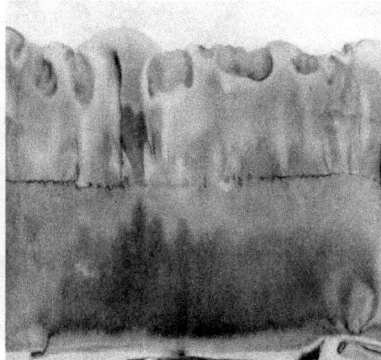

Equisetum tea: fresh (left); fermented (right)

We have been using and recommending the fermented Equisetum tea since the late 1980s. Furthermore, based on Steiner's statement that we "sprinkle it as liquid manure over the fields," as well as practical application in gardens and pastures over several years, the fermented tea BD508 applied as a soil spray following BD500 in the spring has proven more effective than fresh tea as a foliar. The preventative role of Equisetum is emphasized with the early season application, although the fermented tea is every bit as effective as the fresh tea when combating fungus that has already appeared.

508: Revealing the Hidden Forces of Equisetum arvense

We are inclined to encourage a late fall application to the soil as well. It should be noted that in our experience the "characteristic smell" mentioned by Kolisko is a rather powerful "rotten egg" or hydrogen sulfide smell, which is every bit as powerful as is evident in a fermented stinging nettle brew or liquor. It is most interesting to realize that sulfur is a major anti-fungal agent in orchard and vineyard practice, so the sulfurous character of the fermented version of Equisetum may bring to bear another dimension of anti-fungal forces beyond the level achieved by the fresh tea.

Equisetum in the Context of the Cosmic Forces in Our Solar System

Steiner frequently mentions "Cosmic Forces" throughout lecture 6. One exercise I have pursued in trying to pay attention to Steiner's admonishment to view the various preparations "not as substances, but as forces" is to imagine which preparation relates to which member of our solar system. My own scheme may differ to a considerable extent from relationships postulated by others. For instance, compared to Lievegoed in "The Working of the Planets and the Life Processes in Man and Earth," my interpretation may also be seen as somewhat simplistic, but it has worked very well for me, due in large part to my previous study of the field of traditional astrology, and my current interest in astrosophy. Of the nine preparations, it is very easy to identify BD500 with the Earth, and BD501 with the Sun. Almost everyone who has approached this question assigns the Yarrow Preparation, BD502, to the planet Venus; the Chamomile Preparation, BD503, to Mercury; and the Dandelion Preparation, BD506, to the planet Jupiter. With respect to the remaining principal members of the solar system, Moon, Mars, and Saturn, my perception differs substantially from other views that have been expressed. My scheme assigns the Moon to the Oak Bark Preparation, BD505; Mars to the Stinging Nettle Preparation, BD504; and finally, Saturn, the outermost planet of the visible solar system, to the Valerian Preparation, BD507.

Equisetum is an ancient genus and the only surviving representative from the class of plants known as Sphenopsida. *Here is an artist's rendering of a prehistoric Carboniferous forest (approx. 300 million years ago) where the Sphenophytes dominated the understory, growing an estimated 60 feet tall.*

508: Revealing the Hidden Forces of Equisetum arvense

At first glance, one can presume that the solar system assignments are now essentially complete, and need not concern oneself with Equisetum as a *bona fide* biodynamic preparation, if in fact, it is merely something confined to use as an anti-fungal agent. However, in my reading of other lecture cycles by Steiner, I found a number of intriguing concepts and images. In the first place, Steiner identifies the Lemurian age as the time of the separation of the Moon from the Earth, and also describes the Earth of Lemurian times as one of highly active and flexible growth forces, especially prior to that separation. Geologically, the Equisetum plant family was extremely abundant during this same age, which is correlated with the Mesozoic period by Eugen Kolisko.[6]

Numerous fossils of Equisetum plants are found in seams of coal throughout the world, and such plants have been determined to have grown to heights of eighty feet or more, and fossil remains have been found in as many as seventy-six layers, one on top of the other, with clay or shale layers interspersed. Such a finding serves to confirm the possibility of the highly active growth forces mentioned by Steiner.

The Lemurian age is also identified by Steiner as a recapitulation of the "Old Moon" incarnation of the Earth, which immediately preceded our present Earth. In further descriptions of Lemurian times, Steiner assigns a very prominent member group of our solar system, which he identifies as the Comet, as belonging to this age. Steiner states that a Comet is born out of the Sun, and dies into the Sun. The Sun can thus be identified with a cosmic/silica force (BD501) from the center of the solar system, while the Comet can be identified with a cosmic/silica force (BD508) from the periphery of the solar system. Very specifically, Steiner also describes the Comet as the bearer of the "primal feminine force," which opposes the "primal masculine force" of the Moon. When one examines the form of the Comet and its pattern of periodic return and then turns to Equisetum arvense and takes

---

6 Eugen Kolisko, "Lemuria: The Mesozoic Age," *Geology* no. 3 (Kolisko Archive, 1998).

a close look at its form, one can see in the physical expression of this plant the same pattern of periodicity in the leaf or frond whorl intervals that are expressed in the Comet. When one further understands that Steiner advises the use of Equisetum as the carrier of a counter-force to Moon forces and assigns to the Moon a "primal masculine" character, it is not out of order to understand the Equisetum plant as a carrier of "primal feminine" forces. What can better serve to harness or balance "primal masculine" (Moon/Watery) forces than a "primal feminine" (Comet/Equisetum) force?

Thus, if/when the Moon force is either too strong or too weak, Equisetum is the appropriate remedy. The key to the timing of application of Equisetum lies in taking note of whether conditions are too wet (Moon force too strong), or whether conditions are too dry (Moon force too weak). In the former case, one is more likely to achieve the desired results if the application is made when the Moon is in a Fire or Fruit constellation. In the case of a drought, if one makes the application when the Moon is in a Water or Leaf constellation, the probability of some moisture relief is much more likely.

Practical Application of an Abstract Intellectual Theory

All of the theoretical considerations in the previous section would remain only so much abstract intellectualism unless they are put to practical use with dependable consistency of results. Initially, the theory was put to practical use when I tried to address the problem of a drought in 1988 with the thought that, if biodynamic agriculture can provide the environment of the plant or the crop with what is needed for the best growth, then applying the right preparation or preparations in a given situation could stimulate the desired results. Since I was not sufficiently astute or clairvoyant to be able to focus on just the right preparation, I resorted to a "shotgun" approach and came up with the "Sequential Spray" followed by BD500, BD501, and BD508 in a very short period of time. The initial result was at least a temporary relief from the drought. Even though I had not reached the point in my understanding of applying in only a Leaf/Water period to stimulate moisture, coincidentally, it appeared that my initial effort was actually applied when the Moon was in a Leaf/Water constellation. Subsequent applications that year, although resulting in rain, did not yield as dramatic a moisture relief because at least some of the applications were made during times when the Moon was in a Fruit/Fire constellation.

A few years later, when my understanding had matured somewhat, during another drought in the early 1990s, when the then governor of Virginia, Douglas Wilder, made plans to suspend or cancel the deer season because of the extreme fire danger from careless hunters in the woods and forests, a concerted group effort was initiated to stimulate moisture. We were able to coordinate a sequential spray application by six different people at seven different sites, involving farm and garden situations, in both Virginia and North Carolina. The rains came very shortly after this application, and the threatened cancellation or suspension of the deer season was no longer necessary. Since that time, many people have employed the Sequential Spray Technique with a high percentage of attracting beneficial moisture conditions.

If nothing else, this technique provides evidence that Equisetum should be regarded as a multi-dimensional preparation, rather than as an afterthought by Steiner, limited merely to countering some fungus conditions. In keeping with an effort to broaden the horizon in using Equisetum, I have chosen to use it in the compost pile as though it were a seventh compost preparation. The observed result is that the moisture distribution within the pile tends to be more even throughout with no cold, wet spots or hot spots exhibiting so-called fire-fanging. Such a practice can be readily recommended for adoption in any composting situation.

Another avenue that reveals Equisetum as deserving greater attention and usage is the incredibly significant increase in the count of beneficial fungi in compost teas when it is added to the tea brewer as a standard practice.

Altogether, there is a great deal more to be explored before we have fully fathomed the potential role of Equisetum in our biodynamic practice. The one caveat I would mention is that all of the recommendations that Steiner makes throughout lecture 6 in the way of weed, insect, and animal pest "peppers" will provide better results if one has previously applied the biodynamic preparations in one form or another to the area of concern before treating with any particular pepper. In particular, it is of prime importance that the Equisetum preparation is applied in order to achieve the optimum results.

Equisetum as Primary Support for "Peppering" and for the Farm Individuality

When we consider various insect and disease pest problems, I would speculate that we will see differences in such problems depending on whether the Moon forces are too strong or too weak. We have a particular movement of the Moon each month, which gives us possibilities of a wide difference in the strength of the Moon's forces. When the Moon is far away from the Earth at apogee, the Moon forces are weaker and when the Moon is quite close at perigee, we have a period when the Moon forces are particularly strong. Closer observation of prevalent insect problems, for instance, might reveal that one would see more flying insects when the Moon forces are weak, and when Moon forces are stronger, we would see more nematode and fungal problems. Orchestrating the particular insect energy that manifests at any given time would be the Moon's sojourn through the different constellations. Maria Thun's work seems to suggest that each weed appears to have a relationship to a particular constellation as the Moon travels through the zodiac.[7]

When we have to deal with the pestiferous members of the higher animal kingdom, Steiner directs our attention to additional members of the solar system that he states have to do with the forces of

---

[7] Maria Thun and Matthias Thun, *The North American Biodynamic Sowing and Planting Calendar 2010*.

508: Revealing the Hidden Forces of Equisetum arvense

reproduction, namely Venus and Mercury, and most especially Venus as it traverses the constellation Scorpio/Scorpion.

To substantiate the caveat of the previous section with regard to our ability to obtain better results of our "peppering" efforts if the biodynamic preparations are first used, I would call attention to several additional quotes from Steiner. In lecture 3, page 55, he mentions, "The *silica* nature is the universal *sense* within the earthly realm." Earlier, however, on page 38 of lecture 2, he requests that we, "Look at the Equisetum plant. It has the peculiarity: it draws the cosmic nature to itself; it permeates itself with the siliceous nature. It contains no less than 90 percent of silicic acid. In equisetum the cosmic is present, so to speak, in very great excess, yet in such a way that it does not go upward and reveal itself in the flower but betrays its presence in the growth of the lower parts."

Finally, on page 118 of lecture 6, as he is about to give us the remedy based on Equisetum, which we now refer to as BD508, he states, "The true force of fertility depends upon the Moon's influence being normal." The several rhythms and movements of the Moon (that is, full moon, new moon, apogee, perigee, descending, ascending, and so forth) provide a full range of possibilities as to strong or weak effects. When further modified by the forces of the particular constellation through which the Moon is traversing, one can be overwhelmed by all the data that one might need to take into account. The agent for "normalizing" the Moon's influence is none other than Equisetum, which through its silica content brings to bear the Cosmic within the farm individuality.

So, not only are the Earthly forces reinforced by an unbridled Moon at play, but the Cosmic forces are there, as well. It is BD501, and especially BD508 (by virtue of its "normalization" of the Moon's Earthly energies), that stimulates the sense organs of the farm individuality. Without that stimulation, and with the varying and often abnormal Moon forces, the other members of the solar system cannot bring to bear their powers within that organ of the farm to which they most relate. For instance, the planet Jupiter cannot bring to bear the forces that relate to the liver in the human being. Thus, the cleansing "liver forces" within the farm individuality cannot fully function if the Moon forces are not "normalized." Such "normalization" is provided to the farm organism only by the use of Equisetum. Similarly, the other planets cannot exert their forces to the maximum for lung (Venus), digestive organs (Mercury), heart (Sun), and other organ functions relating to Mars, and Saturn.

When the biodynamic practitioner relegates Equisetum merely to the more insignificant role of fungus prevention, the full expression of the farm individuality is hindered. It is this picture of the importance of Equisetum that is so compelling to me and that drives my insistence on the absolute importance of using *Equisetum arvense*, Horsetail herb, (BD508) in our biodynamic practice.

Pest "Peppers" and Homeopathic Dilution

Hugh J. Courtney
Applied Biodynamics, *nos. 27 & 28 1999–2000*

Part I

In the little-understood lecture 6 of *Agriculture: Spiritual Foundations for the Renewal of Agriculture,* Rudolf Steiner speaks of how the biodynamic farmer or gardener may deal with the subject of pests, be they weed, animal, or insect. In this initial article on the subject, I would like to deal first of all with what I perceive as a prerequisite to working with any of the categories of pests, and for the remainder of this article, I would like to limit my discussion to the handling of weeds only. Subsequent articles will attempt to address insect and animal pests in the specific detail required.

I would like to suggest that no effort at countering pests in any category can be successful to its maximum extent if one does not first pay very close attention to what Steiner says regarding the Moon forces and the means he suggests to counter the impulse of those Moon forces when they are out of balance.

Let us first examine what Steiner says about the Moon's forces in general. Near the beginning of lecture 6, Steiner delineates those forces as follows:

> It is common knowledge that the surface of the Moon reflects the rays of the Sun, directing them back toward the Earth. We see these reflected rays of the Sun because we catch them with our eyes and the Earth catches them, too. The rays of the Moon are reflected Sun-rays, but the Moon has imbued them with its own forces, and so they strike the Earth as lunar forces and have

been doing so ever since the Moon separated from the Earth. This lunar force from the cosmos has an intensifying effect on everything earthly....

As I said before, we usually imagine that the Moon simply takes up the rays of the Sun and reflects them onto the Earth. In other words, when we consider the effects of the Moon, we usually think only of the sunlight. But that is not the only thing that comes toward the Earth. Along with the moon beams, the entire reflected cosmos comes toward the Earth. The Moon reflects everything that comes toward it. In a certain sense the whole starry heavens are reflected by the Moon and stream toward the Earth, although you couldn't prove it by any physical means available at present. It is indeed a very powerful cosmic organizing face that radiates down from the Moon into the plants, so that the plants are also enabled to form seeds, so that the force of growth can be enhanced to become the force of reproduction.[1]

My interpretation of these statements by Steiner leads me to see his description as a situation wherein the Moon totally appropriates to its own purposes not only the sunlight (and sun forces), but also the light (and forces) from all the planets of the solar system, and beyond that all the light and forces from the rest of the cosmos, and in so doing converts them as it were to "intensified earthly forces." In other words, it transforms these forces and makes them almost totally its own, by imbuing them with lunar quality, which, having once been joined to the earth, is able to stimulate an enhanced earthly force. It is this lunar quality raying down upon the Earth that leads to strengthening the relatively weak growth forces available within the Earth alone to the point where reproduction is possible for the plant world.

Steiner suggests that some plants also require additional support from the cosmic forces raying down from Mercury and Venus as well.[2] One can speculate that these might be plants with a strong medicinal quality or with other unusual properties such as the Venus

---

1   Steiner, *Agriculture*, p. 116–117.
2   Ibid., p. 117.

Pest "Peppers" and Homeopathic Dilution

Fly-trap with its quality of mobile behavior. In the case of Moon, Mercury, and Venus, we are dealing not only with their specific lunar, Venusian, or Mercurial forces but also what they each collect from the rest of the solar system and the rest of the cosmos. One should also keep in mind a description by Steiner that appears much earlier in the lectures, specifically in lecture 2, where he says:

> A lively interchange is constantly taking place between what is above ground and what is below ground. In addition, we can say that the above ground activity is directly dependent on the Moon, Mercury, and Venus as they support and modify the Sun.[3]

Elsewhere in this same lecture Steiner speaks of air and warmth above ground being "dead" and water and earthly matter as having "greatest vitality" above ground. Indeed, in the living plant, we have a prime example of earthly matter and watery substance with a greater degree of vitality. Conversely, water and earthly matter below ground are dead, whereas air and warmth below ground are enlivened. The mineralized quality of a plant root below ground further illustrates this with respect to watery and earthly substance. Thus, above the soil surface, Moon, Venus, and Mercury forces are working in water and vital earthly substance. Insofar as plant matter above ground is composed largely of water and mineral/earthly substance, it is here that these forces can hold sway to the greatest extent. Indeed, it is just in this above-ground zone that the reproductive enhancement takes place that is the sphere of activity of these planetary bodies. The major part of the reproductive enhancement takes place through lunar forces, which work especially through the element of the living water above the ground.

It is just this reproductive enhancement that is at work when we have a serious weed problem. As Steiner states:

---

3   Ibid., p. 29.

We would get the best stand of weeds if we were simply to let the beneficent Moon have its effect on them and did not hinder it in any way. The weeds would reproduce and multiply, especially in wet years when the lunar forces work best. So, we can conclude that in a wet year, we have a good chance of a more serious weed problem.[4]

Steiner then goes on to say that the key to the weed problem is to somehow "prevent the Moon from exerting its influence on the weeds." It is interesting that, although he makes the strange statement that we cannot "simply switch off the Moon," he tells us that we must "treat the soil so that it will become unsuitable for absorbing lunar influences. And not only the soil, but also the plants."[5] What Steiner next describes is the method of ashing the weed seeds, and using the resultant ash to treat the soil so that the plants or weeds will be reluctant to grow in the soil. Thus, the ashing process is the means whereby the plants are less able to absorb the lunar forces. Later on in this article, we will cover in much more specific detail the precise methods and techniques for the ashing process and for the treatment of the soil for weed problems.

For now, I would like to return to the suggestion by Steiner that we need to treat the soil so that it will be less suitable for absorbing the lunar forces and so that the weeds (and possibly other organisms?) will be "influenced only from outside by the direct non-lunar forces," so as to curb the proliferation of the weeds. Is there, in fact, a way in which we can switch off the lunar forces? I submit that there is such a way, and even though it appears that Steiner is making a humorous comment here, in fact, he is preparing us for the method that he will later invoke in which we can switch off the too-strong Moon forces. It is during unusually wet years when one can expect the lunar forces to be too strong.

---

4   Ibid., p. 118.
5   Ibid.

Pest "Peppers" and Homeopathic Dilution

So what happens when we have a very seriously wet year, along with other conditions that can use the lunar forces to be even stronger than normal? Some of those other conditions could be during periods of a Full Moon, when the Moon is at perigee, or when the Moon is in a watery constellation, perhaps with support from other planets (especially Venus or Mercury?) in watery constellations. In this instance, the lunar forces begin to work into the mineral/earthly matter of the plant and we have the conditions for fungus and mildew. Steiner's description of this situation is as follows:

> Let us assume, however, that the Moon's influence is too strong, that the soil is overly enlivened. In this case, the vitality works up too strongly from below, and something that should occur only in seed formation starts to happen earlier. When the vitality is too strong, it doesn't reach all the way to the top; its very intensity makes it start working lower down. Thus because of the effect of the Moon, there is insufficient force for seed formation. The seed incorporates a kind of dying life into itself, and through this dying life a kind of second ground-level is formed above the level of the soil. Although there is no actual soil up there, the same influences are present. As a result, the seed or upper part of the plant becomes a kind of soil for other organisms. Parasites and all kinds of fungi appear—blights and mildews and the like. What wants to work upward out of the soil is kept from reaching the right height because of the overly strong lunar force. It is remarkable that this happens when the lunar forces are too strong rather than too weak, but that's the way it is. A healthy seed-forming capability is absolutely dependent on the lunar forces being normal rather than too strong. Theorizing and speculation, rather than perception, might lead to the opposite conclusion, but that would be wrong. Direct perception reveals what I have just described.[6]

What "wants to work upward out of the soil" and cannot because of the strong lunar forces is the cosmic/silica force that Steiner associates especially with the outer planets Mars, Jupiter and Saturn.

---

6   Ibid., p. 118.

Steiner's remedy for the overly strong lunar force is to use liquid manure made of a fairly concentrated tea of the horsetail herb, *Equisetum arvense*, referred to as BD508 in some instances. This tea is then further diluted before application and used in a kind of homeopathic manner. Steiner describes this herbal tea application as "simply the opposite of the process I described earlier."[7] I take the process described earlier to mean the entire paragraph wherein Steiner speaks of the overly strong moon forces, just before he gives us the fungal remedy of horsetail tea. If, in fact, the horsetail tea application is an opposite process from Moon forces deemed too strong, might it not also be capable of generating an opposite force when these same Moon forces are too weak? It is this radical thought that, in 1989, prompted my first use of the technique that I subsequently termed the sequential spray technique, which proved surprisingly efficacious in ending the drought conditions we were experiencing at that time, and, when properly applied, has also worked for many others. At this time, I would also like to draw attention to the fact that Steiner's description of the tea uses the term *manure* in association with the horsetail tea to be used.[8] The only early biodynamic practitioner who gives a recipe for a manuring process in conjunction with the horsetail tea is Lily Kolisko in *Agriculture of Tomorrow* when she suggests that the tea be prepared and then allowed to ferment.[9]

With respect to the "radical" thought in the preceding paragraph concerning the insufficient moon forces and the ability of the horsetail herb tea to act in a manner opposite to those moon forces, another of Steiner's lectures also had a serious influence on shaping the conclusions I have reached concerning the possible uses of the horsetail herb. Unfortunately, up to this point, I have not managed to relocate the exact citation. In the unidentified lecture, Steiner speaks of the influence of comets, and amazingly, identifies comets first of all as a

---

[7] Ibid., p. 128.
[8] See *Applied Biodynamics,* issues 6 and 25.
[9] Kolisko, *Agriculture of Tomorrow,* p. 85.

carryover from the "Old Moon" stage of the Earth's evolution, and then, in the present stage of Earth evolution, as embodying a "primal feminine force," which is the opposite of the "primal masculine force" of the moon.

My first reading of this characterization by Steiner immediately prompted a memory of Steiner's statements in the Agriculture Course regarding the application of horsetail herb tea as being an opposite process to the too strong moon forces. Then, a mental picture of the horsetail plant juxtaposed with an image of a comet and its periodic return led me to see the horsetail plant with its leaf structure expressing itself periodically at each node as a visible analogy to the heavenly comet. The fact that the horsetail is also judged to be among the most ancient of plants merely served to reinforce its strong relationship to the moon in my mind. Each reader will need to decide whether the gyrations that my thought processes went through to establish this connection among the horsetail herb, the moon, and comets, are too much of a stretch.

Nevertheless, in claiming such a connection, one must assign a much greater level of importance to the use of the horsetail herb in biodynamic practice. Much of European practice ignores the horsetail herb, and some Europeans have expressed dismay and have even made strong objections to identifying this as a preparation. Among Europeans, only Maria Thun makes frequent suggestions for using the Equisetum in a regular and consistent manner. Those who have attempted to verify her work with lunar and other celestial influences and have failed to arrive at her same results have also likely failed to use Equisetum in their efforts. Since much of the European attitude toward horsetail is unsympathetic and even downright contemptuous, it is not surprising that it would not have been taken into account in attempting to repeat Maria Thun's work with the lunar rhythms of fruit, root, flower, and leaf. However, it is my belief that their failure to do so means that the only forces at play in their biodynamic efforts are those lunar forces that strengthen the earthly. It is only when one has used BD508, thereby taming the primal masculine force of the moon that the cosmic forces of the other members of the solar system can fully enter into the rhythm of plant growth.

For this reason, I would contend that until one begins to incorporate the horsetail herb, *Equisetum arvense*, or BD508, into their biodynamic practice on a basis that regards it as equal in importance to any and every other one of the biodynamic preparations, one is not fully biodynamic, nor will one be able to grow plants that are fully receptive to the forces of all the members of the solar system. Plants grown without due and proper attentiveness to BD508 will be less receptive to cosmic forces and will continue to be influenced mostly by the lunar rhythms of the new and full moons just as is the case with chemically and organically treated plants. Indeed, it is just the same with plants out in nature, when support from a full course of biodynamic preparations including BD508 is not available. In very simple terms, the horsetail herb serves most emphatically to regulate the watery forces through which the lunar forces operate so strongly. Leaving it out leaves us functioning at a considerably less effective level in our biodynamic practice.

Therefore, I would say that effective weed and other pest control could only begin by using BD508 in order to switch off the Moon (and sometimes to switch it on, as well). When Equisetum has taken its place as a fully acceptable biodynamic preparation, then our efforts at biodynamic growing will be much more fully supported by the cosmos.

With a case now made for the value of horsetail as a full-fledged biodynamic preparation that fully deserves to be called BD508, let us now turn to the problem of making and applying a weed pepper. Through the years during which biodynamics has been practiced on the earth, a number of people have worked with the process, but very little has been published on the subject for a variety of reasons, mostly because so few practitioners have taken the time to write about it. In recent years, one biodynamic practitioner and researcher, Maria

Thun, has done a considerable bit of work with the ashing process and has published some of her work. Regrettably, it has not yet been translated into English and is available only in German under the title *Unkraut* (Pests). A few paragraphs on the subject can be found in her planting calendar, *Working with the Stars*, over the last several years; however, the data and photographs depicting her work available in *Unkraut* have been left out. There also appears a brief reference to dealing with weeds in Maria Thun's *Work on the Land and the Constellations*, but it is not of direct practical help in describing the methodology of ashing weeds. One can also find some references to the ashing question in the recently published memoirs of Count Adalbert Keyserlingk, the son of Count Carl Keyserlingk who hosted the Agriculture Course at Koberwitz. One can also find some mention of ashing in Kolisko's *Agriculture of Tomorrow*.

In an effort to aid the biodynamic practitioner, I will attempt to describe a detailed methodology of ashing of weeds based on a synthesis gleaned from all of the publications mentioned in the preceding paragraph, as well as a few other published references, which I cannot necessarily cite. I would like to start with a lengthy excerpt from *Agriculture of Tomorrow* since that publication has now been out of print for some time, and may not be accessible to the majority of our readers at this time:

> We must bring something into the soil that the weeds do not like. It is very simple but effective advice that Rudolf Steiner gives us. The power to reproduce new plants is concentrated in the seeds. We burn the seeds of those weeds, which we want to restrict in their growth, collect the ash, and scatter it over the field. We do not need a great quantity of seeds. The ash radiates out over a large area, counteracting the reproductive force the seeds contain.
>
> We carried out experiments, for instance, with seeds of thistles. After two years we still found thistles growing on the spot where we had used the preparation of burnt thistle seeds. The plants looked quite healthy, but when the seeds began to ripen, we noticed that they began to deteriorate. In the third year, the

thistles were very scarce, and after four years of treatment, they disappeared completely from the treated area.

Nearly all our experiments with weeds have shown us first the phenomenon of seed-deterioration, and the plant no longer grew in the treated district. The whole process takes about four years, and each year the treatment has to be repeated. This again is a very economical and healthy way of disposing of unwanted plants. We do not need chemicals, which damage the soil or poisons to kill the weeds. The most natural thing is done: we combat the plant with the counterforce of the same plant.[10]

Koliško gives us the theory or principles behind the concept of ashing. In the next excerpt, headed "Some practical hints for making these preparations," we are given a brief sketch of how to go about actually doing the ashing process:

The seeds of all the different weeds we want to get rid of are first collected. Of course, they must be quite ripe. Then they are burnt in the open on a small heap of wood, the remaining ashes are collected (wood ash plus burnt seeds) mixed in a mortar with a pestle or some other suitable tool, and then the ash is scattered over the field.

Or we can burn the seeds in a frying-pan and get only the ash of the seeds without the burnt wood. In order to distribute the ash over the required area, we take some potting sand, or soil from the field, and rub it together with the burned seeds very carefully, as we described in the chapter "Smallest Entities," for remedies that are insoluble in water and are potentized with help of a neutral medium like lactic sugar. So we potentize the burned seeds with the soil, and scatter this potency over the field in which the weeds are growing. Every farmer and gardener can easily make these preparations. They do not cost a single penny, and definitely help him to get rid of unwanted plants.[11]

When we more closely at Maria Thun's work, we find a number of significant suggestions for combating weeds that go well beyond

---

10   Ibid., p. 241.
11   Ibid.

Pest "Peppers" and Homeopathic Dilution

what is mentioned by Kolisko in the material quoted above. In the first place, Maria Thun found that each particular weed appeared to germinate best when the Moon was in a particular constellation, for instance, pennycress (*Thlaspi arvense*) when the Moon is in Aquarius/Water Bearer, or chickweed when Moon is in Gemini/Twins. It would be important, however, for each grower to conduct his/her own experiments, since each geographic location may have individualized conditions that might cause the Moon's position in a particular constellation to stimulate a different group of weeds to germinate in a different area of the world. If the grower were able to dedicate an area of land where one could cultivate approximately equal areas, either on a daily basis or at least within the first few hours after the Moon moved into a new constellation, one could determine the weed germinating relationship to the Moon's position in the constellations for his/her own particular geographic location.

The most helpful discovery made by Maria Thun was that when the soil was worked during periods when the Moon was in Leo/Lion, one had abundant and varied weed germination; whereas working the soil when the Moon was in Capricorn/Goat resulted in a minimum of weed germination. Thus if one cultivates the soil when the Moon is in Leo/Lion, thereby stimulating many types of weeds to germinate, and then follows with another cultivation when the Moon is in Capricorn/Goat, the weed seedlings will be disrupted in their growth and end up contributing to soil improvement as a green manure of sorts.

Thun particularly recommends that for potatoes and root crops the last soil preparation and tillage should be done when the Moon is in Capricorn/Goat. In Maria Thun's work, she found that if only the weed ash itself was used, it was important to apply the ash in accordance with the Moon's constellational position when that particular weed was at its highest germination capability. Thun found that the single most effective step that one could take in using the ashing technique was to *dynamize* the seed and wood ash together.

Dynamization is described as grinding the products of the ashing process using a mortar and pestle for a period of one hour. The second most effective step to be taken is to use a homeopathic D-8 potency as the final product to be sprayed on the area one wishes to treat. The subject of potentization and arriving at a D-8 potency will be disclosed in Part 2 on this subject.

An additional weed control technique mentioned by Maria Thun involves making liquid manure of weeds that are not particularly suited for a compost heap, especially those with runners or creeping habit in their growth. Such weeds as ranunculus, couch grass, thistles. goutweed, coltsfoot, chickweed, and some members of the mint family are better dealt with by using them to make liquid plant manure. This is accomplished by filling a barrel with water and putting the weeds of this nature in a goodly quantity into the water. On leaf days it is recommended to stir the liquid for a few minutes. When all the plant matter has completely deteriorated in the water, the resultant liquid can be sprayed when the Moon is in Cancer/Crab (the constellational position for the Moon assigned to these weeds by Maria Thun). The area from which the weeds were taken, or where they are not wanted is then sprayed three times. With repeated sprayings, these troublesome weeds will be encouraged to disappear. These particular weeds also promote good growth in cruciferae, cucumbers, and tomatoes when applied as a highly diluted liquid manure, and can even be poured over a compost pile in order to improve the growth factors for the compost itself.

A further suggestion made by Maria Thun for more effective weed control or rather for prevention of future weed problems is to avoid applying products to the soil that are derived from an animal substance such as horn, bones, blood, bristles, wool, feathers, or meal taken from the bodies of animals. All such substances should be composted with manure, fresh grass, or plant material. These materials originating from the bodily substance of animals should only be applied as finished compost in order not to promote weed growth

Pest "Peppers" and Homeopathic Dilution

(please note that Demeter standards now prohibit the use of blood and bone meal even in compost).

One final reference as background to the consideration of the ashing process comes from a book newly available in English by Count Adalbert Keyserlingk, the son of the host and hostess of the *Agriculture* course given in their home at Koberwitz. In his memoirs titled *Developing Biodynamic Agriculture*, Count Keyserlingk describes some of his ashing experiences as follows:

> A whole bucketful of charlock seed was burned for one field experiment. Before that, we had already established that the seeds must not remain in an incandescent state for long in the ashing process. The temperature had to be kept relatively low so that the incandescent phase was missed out and a browny ash would remain. Years earlier I had used the process on chicken farms in England and found that the power that is the opposite of the germinating power is lost at red heat.[12]

This is perhaps a fairly important factor to take into account, inasmuch as some methods of ashing would quite likely involve reaching an incandescent state rather readily: for instance, burning the seeds in a frying pan covered with a lid and placed in an oven or in a wood-burning stove or fireplace, and then leaving them for a long time. Some individuals have claimed that effective results were obtained in such situations when the resultant seeds were essentially turned to charcoal by the ashing process, although the individual seeds did retain their form when the process was completed.

First of all, of course, one must harvest the weed seeds. Here, we immediately have the question of whether one must address each individual weed separately or whether one can address a number of problem weeds at the same time. Here at JPI, we have been taking each weed individually because of our fondly held hope that we shall be able eventually to develop a library of various weed peppers, which would be available to the biodynamic practitioner whenever the need

---

12 Keyserlingk, *Developing Biodynamic Agriculture*, p. 38.

arose. For most garden and farm situations, however, a common fire for a number of different weed seeds might be the most expeditious route to take. In any case, one should gather a goodly number of fully ripe weed seeds. If one is merely treating a large garden area, then a teaspoon or a tablespoon full of each seed may be quite sufficient. On the other hand, for many acres of crop fields, one may wish to have a sizable bucket full of weed seeds for which control is sought. While there are any number of methods for putting the seeds through the fire, some of which have already been mentioned in the previous references, I will confine my description to the technique we have employed at JPI for the major part of our ashing experiments.

We use a metal tubular device that is sold through any well-stocked hardware store for the purpose of igniting charcoal without using lighter fluid. This charcoal starter has a metal platform about one-fourth of the way up the tube from the bottom. Since this platform has very large holes in it, which would not allow us to build a proper fire for ashing the seeds, we cut a circle of hardware cloth to fit exactly on this platform. We prefer to use hardware cloth of one inch, although for larger seeds or for ashing of some insects as well as most animal skins, one could use hardware cloth of one-quarter inch mesh. On top of the hardware cloth supported by the metal platform, we build a wood fire. The source of wood for this fire is oak bark since we have a goodly supply of thick oak bark pieces taken from already timbered trees by someone trying to help us out with the oak bark preparation without realizing we much preferred to take the oak bark from living trees and to keep only the outermost bark when harvesting material for this preparation.

Pest "Peppers" and Homeopathic Dilution

Any good firewood would be suitable, but it should be reduced to the size of small kindling. In any case, we build a little terrace of the oak bark chunks on top of the hardware clothing on the metal platform; start a fire underneath the pile, using paper, such as newspaper or scrap paper from the office; and then, when a good strong flame is going, we dribble the seeds slowly into the flame.

Underneath the charcoal starter tube, we have a metal plate, which we have used otherwise for crushing quartz. We have also used a very large cast-iron skillet sometimes as well. The plate or skillet catches the seeds and wood ash that have passed through the flame and fallen through the hardware cloth and below the metal platform. When everything has cooled sufficiently, we use a small brush to sweep all of the seed and wood ash residue into a metal bowl.

The next step is the one Maria Thun believes is extremely important, that is, the dynamization process, in which we use a pestle to thoroughly grind all of the seed and oak bark ash residue together for one hour. At this point, one could choose to distribute the resultant powder directly on the soil of the area for which you are seeking protection. However, because of the small quantity of ash usually available, it may be preferable to take the dynamized ash through a potentization process before spraying or sprinkling it out on the soil. Once the final product for such sprinkling is determined, one should plan on three or four applications in a row, to be applied even on the same day according to Maria Thun. One could also choose those three or four applications on the subsequent Moon passages through the preferred constellation in case one is doing only one specific weed.

If we do not have a clear certainty of the proper constellation for either burning or application of the ash, I would recommend a time when the Moon is in a Fire constellation close to Full or New Moon. In keeping with Maria Thun's findings regarding weed germination, the first choice would seem to be when the Moon is in the constellation Leo/Lion. In any case, such applications should be repeated each year for a total period of four years in order to achieve something approaching 100 percent effectiveness as far as desired results. In the second year, the weed(s) in question will show various signs of weakness in growth and will be less likely to set viable seeds. and by the fourth year, they will essentially cease to be a problem. If this should not be the case, then any number of things could have given the poor results. The most probable cause for such a situation would have to do with the timing of both the ashing process and the applications. If one fails to note possible negative aspects, not all of which are fully taken into account with the various biodynamic planting calendars, then one could have inadvertently chosen a poor time either for the ashing or for one of the applications. Failure to note occultations, a planetary node, a wholesale series of conjunctions, or other malefic aspects, could contribute to a lack of concrete results

Above all, one needs to constantly observe the reality of what is taking place in the plant kingdom as a result of any weed peppering that is used. Simply taking someone else's word, including mine, or Maria Thun's, or even Rudolf Steiner's, allows one's biodynamic practice to become rigid and inflexible. When we take what someone else has written as the only methodology to employ, we put biodynamics in a straitjacket and it becomes a dead dogmatic approach to agriculture, rather than the living agriculture that Steiner intended when he encouraged people to experiment for themselves.

What I have presented here is merely one person's effort to come to a better understanding of the lunar forces and of the primary tool Steiner gave us for working not only with those lunar forces, but also to counter pests of all kinds, be they weeds, insects or animals. Examine for yourself my premise as to the value and necessity of using horsetail herb (*Equisetum arvense*) not only for dealing with fungus problems, actual or potential, but also as far as pests are concerned. I believe you will be glad when you place horsetail herb (*Equisetum arvense*) in its well-deserved position as a full-fledged biodynamic preparation, BD508.

Part II

Part two concentrates on "potentizing," or homeopathic dilution. It is concerned specifically with applying the results of our ashing process. As mentioned in part one, once one has finished burning the weed seeds, there is usually only a modest quantity of ash available to apply. To ensure the recommended three or four sprayings in a row, as well as repeated applications during a year or a four-year period for the most effective results, one can potentize the ash and provide enough material for numerous treatments of considerable territory.

To begin, I offer a definition of the potentizing process, and at least some explanation of why it is presumed to work. The words *potentize* and *potentizing* do not seem to appear in most dictionaries. One can find the word *potentiate*, which means, to make more potent. A useful source for understanding the meaning of *potentize* and *potentization* is *The Basis of Potentization Research* by Theodore Schwenk. His most noted publication in English is *Sensitive Chaos* and is best known for his study of water rhythms and movement and especially for his development of the water-drop method of quality testing. To reach a thorough definition, I quote at length from *The Basis of Potentization Research*:

> Potentization, which is a stepwise treatment of the original substances, was introduced by the homeopathic school on the basis of a discovery by Samuel Hahnemann (1755–1843) at the turn of the nineteenth century. Although Hahnemann started out prescribing material medications, he later came in the course of his practice to ever higher potencies. The historical development of his discovery was thus itself just such a series of steps from a material beginning to use of high potencies. However, the essential characteristic of the homeopathic method is not, as is often mistakenly supposed, the producing and prescribing of potentized substances, but rather its adherence to the principle of "like cures like" or, in medical terminology, the "Simile Principle." This discovery actually predated that of the effectiveness of potentized substances. In a life devoted to attentive study of

the effectiveness of medications, Hahnemann had the distinction of adding to the Simile Principle the method of the potentizing process. Though he could not have been aware of it, this process must have been known very early in cultural history, a fact reported by Rudolf Steiner from his spiritual research but of which little other record exists.

In the sketch to follow, potentizing will be discussed chiefly in its aspect as a pharmaceutical process, though it does play a much larger role in nature than is commonly recognized. We remind our readers of the great significance of the trace elements occurring in the very finest dilutions, of the existence of many medicinal springs with their traces of substances important to life, of the ocean tides with their rhythmical motion, their surf breaking along the coasts with its variety of dissolved substances. These processes can all be regarded as nature's own worldwide potentization processes, and this dissolving takes place according to such laws as govern the gaseous element.

Potentizing is done, as is known and has already been mentioned, in a series of steps. A quantity of a given substance is taken up by a quantity, several times larger, of a suitable medium: water, alcohol, milk sugar, and subjected to rhythmical movement. Superficially considered, this corresponds to a dilution; a decimal potency is a plus proportion, a centesimal potency 1 plus 99. Once a proportion has been chosen, it is retained through all the stages of the potentizing process.

But the process involved here is more than a mere diluting. The important process, as Rudolf Steiner once put it, is "what is being made to happen": in the first place, a rhythmical motion at every stage of the process. In his day, Hahnemann ascribed greatest importance to the process involved in treating the substance. "The inner remedial capacity comes marvelously alive when subjected to friction (shaking), and frees itself as it were from its bonds with matter so as to be able to work more freely and with greater penetration on the human organism."

To put it thus is to characterize potentization in its three fundamental aspects of substance, force and rhythmical motion. It is the latter that leads over from the substantial state to the development of force.

Pest "Peppers" and Homeopathic Dilution

With this lengthy explanation of the process, we now have a basis for understanding why potentization may be effective. When one takes the mother substance, dilutes it with a medium of water or milk sugar at a ratio of 1 to 9 or even 1 to 99, and then subjects it to a rhythmical motion, one transfers the force inherent in the mother substance to all the particles of the medium. At the same time, there appears to be a further enhancement or liberation of the force of the mother substance so that it takes on a more powerful or potent quality.

A rough analogy can be found in the process whereby our muscles become stronger when we exercise them regularly with a barbell in a weight lifting program. The process of potentization also will help explain why the label on the bottle of cough syrup or other medicine requests the user to "shake well before using." Whether such a label is placed there consciously or unconsciously, following its direction results in a more effective medicine.

Let us assume a concrete example and walk through the process step by step, and thereby, we hope, answer most, if not all, of your questions on the process of potentization. Please keep in mind that this description of the process will apply not only to weed ashing but also to the ashing of insect or animal pests.

We start with the "dynamized" ash. Assuming that we have on hand approximately six tablespoons of this dynamized ash, we need to choose a measure or weight that will remain consistent throughout the entire process from DØ to D8 (or CØ to C8 if the centesimal dilution route is our choice). For the sake of our example, let 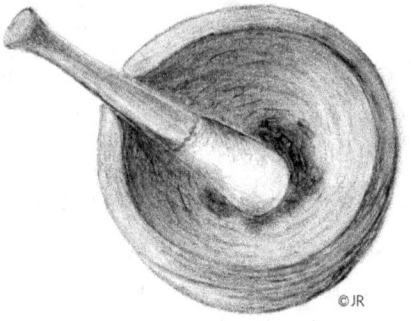 us choose to start with a volume measure of one tablespoon of dynamized ash. Since most biodynamic practitioners do not have a laboratory scale that would allow them to measure out a gram or so of ash, nor calibrated metric measuring containers for millimeters of water, we will stick to the ordinary implements one can find in the kitchen. Of course with a metric laboratory scale and metric flasks, the calculations may be simpler, because one is always dealing in easily calculated decimal proportions. If you intend to do lots of ashing, it may well be worth your while to obtain the necessary metric-based equipment. At any rate, remaining in our ordinary kitchen implement mode, we take our tablespoon of ash, place it in a small jar, add nine tablespoons of water, tighten down the lid thoroughly, and begin a rhythmical motion that will result in a thorough mixing of the ash with the water. The object of the mixing is to have every molecule of the ash slide past every molecule of the water, thereby imparting to the water such forces as are contained in the ash. The required rhythmical motion can be accomplished in a number of different ways and for varying lengths of time, depending upon which homeopathic or anthroposophic medical school or authority you wish to follow.

In homeopathy, the rhythmical motion is achieved by a process termed *succussing,* or *succussion.* One takes the jar or container in one's fist and firmly but gently pounds it against a rubber mat or a book for a specific period of time, or for a specific number of times. Some of the usual recommendations are to continue the rhythmical pounding for a period of one minute, or alternately a period of three minutes. Another recommendation is to continue this rhythmical pounding for a count of 100.

In anthroposophic medicine, the technique of succussion is thought to impart a somewhat violent character to the remedy, so a different technique is employed. One takes the container or jar and holds it in one's hand(s), while moving the jar in a rocking or back and forth, circular motion so that the ash and water in the container begin to move in a figure-eight or lemniscate pattern.

Whichever methodology you employ in your effort to achieve the thorough mixing required, it is suggested that you be consistent

Pest "Peppers" and Homeopathic Dilution

throughout. After completing the "potentization" step with the DØ ash to which you have added measures of water, you now have a D1 potency with a volume of 10 tablespoons, or ½ cup plus 2 tablespoons. Our ultimate target for most ashing treatments is a D8 potency, and we will tabulate the volume required if at each step one takes the entire volume of the preceding step and adds more measures of the diluting medium:

Step	Volume at completion
D1	10 tablespoons or ½ cup + 2 tablespoons
D2	100 tablespoons or 6 ¼ cups or 1 quart + 2 ¼ cups
D3	1,000 tablespoons or 62 ½ cups or 3 gallons + 3 ½ quarts + ½ cup
D4	10,000 tablespoons or 625 cups or 39 gallons + 1 cup
D5	100,000 tablespoons or 6,250 cups or 390 gallons + 2 quarts + ½ cup
D6	1,000,000 tablespoons or 62,500 cups or 3,906 gallons + 1 quart
D7	10,000,000 tablespoons or 625,000 cups or 39,062 gallons + 2 quarts
D8	100,000,000 tablespoons or 6,250,000 cups or 390,625 gallons

Quite obviously, at a fairly early stage, one is dealing with an enormous volume of material that would be impossible to handle by succussion or the lemniscate rocking motions described above. Therefore, one is well advised to do two things before proceeding beyond the D1 potency. The first thing is to determine how much of the finished product of D8 you want to apply. If you were going to treat one acre or less then you would probably need no more than 3 to 5 gallons to spray for your final D8 potency. Thus, at the D7 step, you would require only ½ gallon or less of material. For a larger area, say twelve acres, you would need 36 to 60 gallons for the final D8 potency for spraying, and at D7 would need 3.6 to 6 gallons of material.

Let's assume that we are dealing with only a one-acre or smaller garden situation. We, therefore, need only repeat for the first several potencies the quantities involved in the step from D1 to D2. In other words, for arriving at D3, we take only tablespoons or ½ cup plus 2 tablespoons (that is, the same quantity we arrived at upon completion of a D1 potency) from the quantity of D2 we have made. To this quantity, we add 9 more measures or 90 tablespoons and proceed with our rhythmical mixing to arrive at a finished D3 potency with a total quantity of 100 tablespoons or one quart plus 2.25 cups. Likewise, for each step up to D6 we take the same quantity. At D6 we can take the entire volume of D6, that is, 1 quart plus 2.25 cups and produce 3 gallons plus 3.5 quarts plus ½ cup of D7. In going from D7 to D8, we take ½ gallon of D7 and produce 5 gallons of the finished D8 potency we want to spray. We could also choose to take the entire amount of D7 and end up with 39 gallons plus 1 cup of D8 for spraying larger areas. At the D6 to D7 step and again at the D7 to D8 step, instead of the succussion or lemniscate rocking technique employed for the earlier steps, we take the given quantity plus the added amount of water and stir for one hour in the usual biodynamic fashion. Admittedly, this might be somewhat confusing to follow, so I will place the foregoing description in a tabular form in the hope of arriving at greater clarity for the reader on page 226.

Before one begins all of these potentization steps, it is helpful if one has lined up a complete set of containers to accomplish each step in the entire process. For instance, in going from DØ to D1, a half-pint jelly jar is an appropriate size. In the example above, if we have gallon or ½ gallon size containers, such as apple juice containers, especially the ones with a handle on them, we can use such containers for each step from D2 through D5, and for D6 to D7 and D7 to D8, we can use the customary stirring vessels employed for the regular stirring of the various biodynamic preparations. It is important

Pest "Peppers" and Homeopathic Dilution

From	To	Take
D0	D1	1 tablespoon and add 9 tablespoons of dilutant Entire amount above and add 90 tablespoons or 5 ½ cups + 2 tablespoons (volume when finished = 1 quart + 2 ¼ cups)
D1	D2	Entire amount above and add 90 tablespoons or 5 ½ cups + 2 tablespoons (volume when finished = 1 quart + 2 ¼ cups)
D2	D3	10 tablespoons or ½ cup + 2 tablespoons and add 90 tablespoons or 5½ cups + 2 tablespoons (volume = 1 quart + 2 ¼ cups)
D3	D4	10 tablespoons or ½ cup + 2 tablespoons and add 90 tablespoons or 5 ½ cups + 2 tablespoons (volume = 1 quart + 2 ¼ cups) Entire amount above and add 3 gallons + 2 quarts + ½ cup (volume = 3 gallons + 3 quarts + 2 ½ cups)
D4	D5	½ gallon of above and add 4 ½ gallons (volume = 5 gallons)
D5	D6	Entire amount above and add 3 gallons + 2 quarts + ½ cup (volume = 8 gallons + 3 quarts + 2 ½ cups)
D6	D7	Entire volume of D6, that is, 1 quart plus 2 ¼ cups and produce 3 gallons plus 3.5 quarts plus ½ cup of D7.
D7	D8	½ gallon of above and add 4 ½ gallons (volume = 5 gallons)

or, alternatively:

D7	D8	Entire volume of D7 and add 35 gallons + 2 ½ cups (volume = 39 gallons + 1 cup)

that whatever container is used be sufficiently large to leave plenty of room for the liquid to move freely throughout the container. It is also important that one carefully label each container according to the potency it holds.

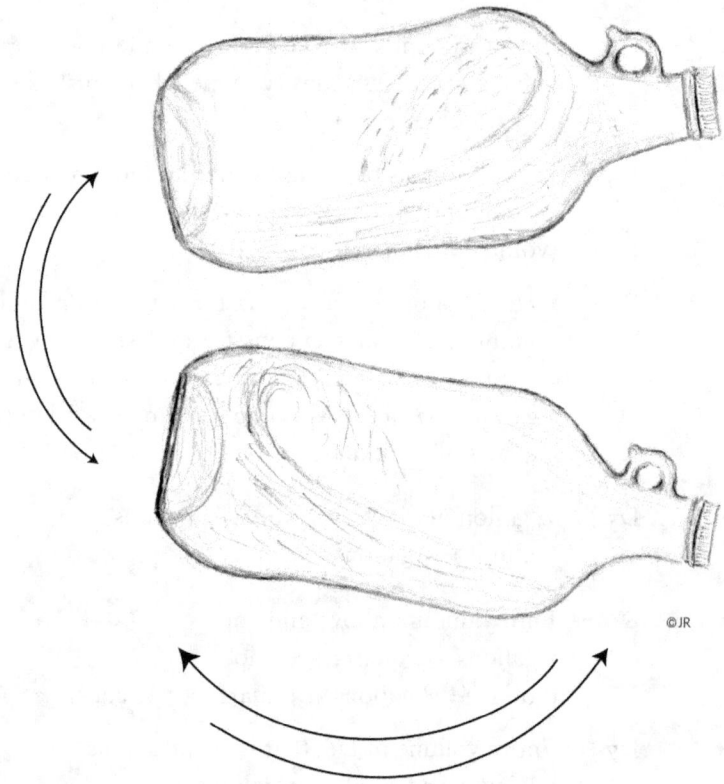

Once you have the quantity of D8 potency desired for spraying, you should try to schedule repeated sprayings, which should not be a problem if you have a series of potencies available for further potentization. For example, the 3 gallons + 3 ½ quarts + ½ cup produced at the D7 step described above for the one-acre situation is sufficient to yield a total of batches of D8 at gallons each. Once all of the D7 has been used up, you will have to return to the D5 step and repeat the D6 and D7 steps. At a certain point, you will need to return to the original ash and repeat each step from the beginning. In homeopathy,

Pest "Peppers" and Homeopathic Dilution

for water-based potencies, there is a question as to the stability of a potency over time, unless twenty percent or more of alcohol is used to stabilize the product.

For myself, I am not comfortable in applying alcohol to soils no matter what the degree of dilution. For a potency in which water alone is used as the medium for diluting, the shelf life or viability cannot be retained beyond a few months, and to be on the safe side, I would suggest that three months is a reasonable time to consider that a potency is still usable. Beyond that time, one should start over again from a solid substance stage, that is, the ash itself. It is for this reason that milk sugar is used to arrive at potencies with a very long shelf life.

It is hoped that the above description is sufficiently detailed and clear and that the reader will be encouraged to potentize the products of the ashing effort with a good deal more assurance that biodynamics also offers a serious and viable solution to various weed and pest problems.

Thanks to Jennifer Reeve for the illustrations for this chapter.

The Three Kings Preparation

Hugh J. Courtney
Applied Biodynamics, *no. 35, 2001–2002*

Adoration of the Magi, Marx Anton Hannas

The Three Kings Preparation

Let us be rightly filled
O Spirit of the World
With spirit embracing mood,
So that we do not fail
To do what heals the earth
And leads to earthly progress,
Wresting from Lucifer and Ahriman
What is right for us.
— RUDOLF STEINER[1]

Given the nature of the times we now live in, where we indeed seem to be in the opening stages of what Steiner termed "The War of All Against All," it behooves us to ask what steps we can take to turn the tide of events so that we can instead experience the "thousand years of peace" promised by the Prince of Peace. The worldwide effort in biodynamic agriculture has not been sufficient to reach the point Steiner so hoped for when he spoke to Ehrenfried Pfeiffer about the importance of seeing that the biodynamic preparations were used widely in the earth. It is probable that he hoped that such would be the case before the end of the century, so that the possibility of the "war of all against all" would be diminished or ameliorated. Such does not seem to have happened, and indeed, there seems to be a certain propensity for internal strife within the biodynamic movement itself. Simply because we practice biodynamic agriculture according to our own individual interpretation of it does not by any means relieve us of the human failing of egocentric and materialistic behavior that is so prevalent these days. Such behavior is a root cause of the "war of all against all," and seems to manifest as readily for some in the anthroposophic and biodynamic movements fathered by Rudolf Steiner as it does among nations in the world.

---

1 Mentioned in Hugo Erbe, *Präparate zur Förderung des elementaren Kräftewirkens im biologisch-dynamischen Land- und Gartenbau* (Preparations for promoting elemental force working in biodynamic farming and gardening).

Facing the fact that the biodynamic preparations have not thus far managed to be used widely throughout the world, are there any other steps that can be taken by those willing to use the preparations beyond what they are already doing? After two year's personal experience, I believe there is indeed something we can do. What can be done is to take up the use of Hugo Erbe's Three Kings Preparation,[2] along with the other nine basic biodynamic preparations that Rudolf Steiner gave us in the Agricultural lectures in Koberwitz in 1924.

Following these thoughts, we include an article that describes the Three Kings Preparation by German-Michael Hahn of Würzburg, Germany. As well, we include the protocol for its making and use as provided by German-Michael who has worked very intensively with this preparation for the last several years.

To begin with, however, I would like to briefly identify who Hugo Erbe was, and then take a more direct look at what Hugo Erbe himself had to say about the Three Kings Preparation. Hugo Erbe was born September 8, 1895, in Bad Cannstatt, Germany, and died October 13, 1965. Born prematurely and rejected by his domineering father, he was brought up by his grandmother. It was his grandmother who lovingly awakened in Hugo a love for and interest in all beings of nature. At the point when Hugo was ready for the world of work, he was initially given commercial training to prepare for entering his father's textile business. This prospect for his future was apparently so distasteful to Hugo that, after a year of such schooling, he volunteered for military service during the war. During his wartime service, while passing an abandoned trench full of trash, he encountered a book amid the debris, which he put into his pocket. Later examination of the book revealed the title, *Die Geheimwissenschaft im Umriss* (*An Outline of Esoteric Science*). From that moment forward, Hugo Erbe became a lifelong student of Rudolf Steiner and anthroposophy.

---

2   Ibid.

The Three Kings Preparation

After the war, Hugo spent a few years in his father's business and had the good fortune to marry Maria Roschmann, the daughter of the owner of a large bakery in Ulm, Germany. Their marriage was an unusually happy one, without any crisis, and produced four children. By 1924, it was no longer possible for Hugo to work in the family business since his father could not tolerate any contradiction of his rulership in his textile kingdom. At this point, Hugo took up a more formal study of singing, having been gifted with a rich natural baritone voice. Destiny, however, had other plans, because just as he was to take his first professional singing engagement with the Frankfurt Opera, he became very ill and lost his voice. Although his illness was ultimately cured, his voice never regained its natural strength. He had, nevertheless, been trained in breath control and tone formation.

He had also gained valuable insights into the human organism as an instrument for forming sounds and words. It was during his illness, as well, that he developed a reverence for the WORD (John 1:1), and added John, the Seer of Patmos, and Christian Rosenkreuz as his spiritual mentors alongside Rudolf Steiner.

After recovering from his illness, Hugo was faced with the need to choose a different career path. A place in the family business was now closed to him because of his difficult relationship with his father, while on the other hand, his illness extinguished his hopes for a career as a singer. Thus it was, at the age of 33, Hugo entered his father-in-law's bakery business. With the intensity so typical of his efforts, he became a dedicated student of this new profession and entered deeply into the being of the grain plant. His studies led him to the development of remarkable new methods of bread making. Using a special combination of honey and pea flour, as well as wine, he created a leavening agent that he ultimately patented in 1937.

His efforts with new recipes generated a substantial increase in the demand for the products of the bakery. The resultant prosperity of the bakery and a café that he opened alongside the bakery allowed Hugo Erbe a certain amount of leisure to travel about the country. His travels with a wartime friend whom he had employed as a chauffeur allowed him to devote considerable time to an observation of nature. He and his friend were able to gather one of the most comprehensive collections of minerals to be found in private hands at that time. Hugo's diligent study of spiritual science and his devoted practice of Rudolf Steiner's spiritual exercises allowed him to build on his childhood awareness of nature spirits. He attributed his success in acquiring a significant collection of minerals to his ability to keep up a good relationship with the gnomes, the earth spirits.

During this same period, Hugo entered into a study of human physiology and the relation of human organs to healing plants, metals, and various mineral substances. He also developed close relationships with several physicians. In addition, he enjoyed friendships with a number of musicians and artists and was acquainted with the author Werner Böhm. Erbe was especially interested in the anthroposophic writings of Böhm, who made an exhaustive study of everything to be found in Rudolf Steiner's work on the subject of the wisdom of the stars. Böhm's work was intended to help with a renewal of astrology and, by using Steiner's insights, Böhm contributed to the developing field of "astrosophy" with the work, *Cosmos, Earth and Human Beings*, which was published in German some five years after his death. Erbe was also acquainted with the forester, Viktor Schauberger. One wonders if Erbe was not the peasant farmer described by Schauberger as stirring clay in a barrel of water while tone-singing before sprinkling his fields with the resulting stirred clay water.[3]

Because of Erbe's forthright and plainspoken opposition to Nazism, he was advised to relocate, and sometime in 1933 or 1934, he moved to a small twenty-acre farm in the region of Markdorf near Lake Constance. It was here at "Gully Farm" that Hugo Erbe

---

3 See Callum Coats, *Living Energies: Viktor Scahuberger's Brilliant Work with Natural Energy Explained*, p. 268.

began a concentrated study of Rudolf Steiner's Agriculture Course and involved himself intensively with gardening and farming.

With his typical intensity and continuing meditative effort, he conducted a great deal of research, both practical and esoteric, which led to the development of entirely new grains derived from wild grasses. His esoteric research, and what he felt was a clear call by higher spiritual beings, also led to the development of an entire constellation of new preparations intended to take Steiner's initial indications to a higher level. With what appeared to his close friends to be well-developed spiritual-scientific insight, rather than intellectual abstraction, Hugo Erbe developed a total of twenty-one new biodynamic preparations as supplements to Steiner's original nine given in the Agriculture Course.

Making and Using the Preparation

What follows is a paraphrase of Hugo Erbe's own description of the making and using of the Three King's Preparation as detailed in the chapter "How to Make the Biodynamic Preparations," which he contributed to the book *Präparate zur Förderung des elementaren Kräftewirkens im biologisch-dynamischen Land- und Gartenbau* (Preparations to promote the elementary forces in biodynamic agriculture and horticulture). The Three Kings Preparation is identified as the eighth of the twenty-one special preparations that Hugo Erbe gifted to biodynamic agriculture. It is one of the two special preparations that Hugo Erbe designated specifically as "sacrifices for the elemental world." First are listed the three primary ingredients in equal portions as follows:

- 1.05 ozs. / 30 gms. Aurum metallicum D_2
- 1.05 ozs. / 30 gms. Frankincense
- 1.05 ozs. / 30 gms. Myrrh gum resin

These three substances are ground together using a porcelain mortar and pestle until a fine powder is produced.

This powder is then thoroughly mixed with 1.75 ozs./50 gms. of rainwater and 1.75ozs./50gms. of vegetable glycerine until an emulsified or colloidal state is reached. This emulsion can be used immediately after preparation, but remains effective for years if stored in an airtight, non-metallic container in a cool, dry, dark area.

For use, about one teaspoon of the preparation is added to approximately two gallons plus two cups of warm rainwater, or sun-soaked pond water. Stir for one hour starting at 1:30 p.m. using a non-metallic container, with a wooden keg or bucket being the first choice for a stirring vessel. The method of stirring is the same as is used for the horn manure or horn silica preparations, that is, stir in the usual biodynamic fashion. It is particularly important in the case of the Three Kings Preparation that the person stirring be fully conscious of and focused on the purpose to be accomplished through the use of this preparation.

After stirring, it is to be sprayed out in the etheric-cosmic inbreathing phase of the late afternoon on January 6, Three Kings Day, also known as Epiphany. The person spraying walks around the perimeter of the area to be protected and stops about every fifty-five yards ± to spray once in the direction of the neighboring fields, that is, aim the spray away from the center of the property to be protected. It takes about ½ gallon ± to cover 3,280 feet of boundary or the perimeter of sixteen acres.[4] In a certain sense, one is forming a magic circle of protection around a particular area.

Erbe specifically termed the Three Kings Preparation a "sacrifice" for the elemental kingdom, and it was to provide protection against the working of opposing powers.

These opposing powers are referred to in Rudolf Steiner's spiritual science as Lucifer and Ahriman, and they stand opposed to the being of Christ. The three gifts of gold, frankincense, and myrrh were brought by the wise men from the orient as sacrificial offerings to

---

4   Please note that the perimeter of one acre is 835 feet ±.

The Three Kings Preparation

the Jesus child in Bethlehem. Erbe states that according to Steiner's representation, gold was brought as a symbol of outer wisdom-filled power; frankincense was a symbol of the sacrificially hovering cosmic ether in which the spirit lives, and myrrh served as a symbol of the victory of life over death. In an additional interpretation by Steiner, gold serves as an image of the wisdom of the past, frankincense represents the transitory nature of the present, and myrrh is a sign of forces pointing toward the future.

Initiates throughout the ages valued these three substances as signs of the awareness of spiritual events that take place behind the physical forms of appearance. The three sacred substances in the Three Kings Preparation serve as sacrificial offerings and as a sign of the working together of spiritual beings in the earthly realm. At the same time, this preparation is a calling upon the divine Trinity and has the effect that elemental beings that have become malevolent feel repulsed and, in withdrawing, lose their power.

The use of this preparation is a request for the help of the spiritual world in our efforts to overcome threatening and harmful powers. By its use, we create a "magical circle" that malevolent beings recognize as such, while at the same time these beings realize that they are recognized for their negative work. Such recognition forces them to retreat from this "magical circle." Erbe experienced repeatedly that malevolent beings such as frost and winter giants cannot penetrate the line made in drawing this magic circle by spraying the Three Kings Preparation.

Erbe felt that a water form was required in providing these three substances as a sacrificial offering rather than a smoke or incense offering that had been the case previously in human history. Thus, the Three Kings Preparation is now available for biodynamic agriculture through the work of Hugo Erbe.

At this point, I would like to quote a few sentences from a letter received from Jennifer Greene dated November 30, 1999. In that letter, which served as my own introduction to the Three Kings Preparation, she states the following:

> Hugo Erbe, a longtime biodynamic farmer in Germany (1895–1965), had a very close connection with the Elemental world. After the bombing of Hiroshima and Nagasaki, he observed a massive disruption and flight of beneficial elemental beings from his farmland. They were being demonized. As healing for the damage done to the organism of the earth, he developed this preparation out of the gifts of the Three Wise Men—the gifts of gold, frankincense, and myrrh prepared as a biodynamic preparation. He discovered that these substances are able to:
>
> - Be an encouraging live elixir for the elemental beings well disposed to us through the gold,
> - Lead the elemental beings back to their favorable connections through the frankincense,
> - Help our higher I to take the lead over bereft and abandoned groups of elemental beings.[5]

In light of the events in the world today, it seems to me that it is increasingly important that more and more people must consciously take up the use of this preparation and thereby express a willingness to work with the elemental kingdom, whether we have a direct awareness of it or not. Humanity has so frequently ignored and betrayed the elemental beings by our desecration of nature. The mere fact that these beings are invisible to the majority of human beings does not mean they do not exist, and the sooner we take some sort of positive action that provides a message to them that we wish to work with them, the sooner we can counter much of the chaos so apparent in the world. No other action on our part can send this message to the elemental kingdom quite as well as our use of the Three Kings Preparation.

---

5 Jennifer Greene is Director of the Water Research Institute of Blue Hill, Maine. She credits Barbara Booth, a Waldorf high school coordinator in Santa Fe, New Mexico, as the one who introduced her to the Three Kings Preparation.

The Three Kings Preparation

There is one quite fundamental warning that I must issue to those who would take up the use of this preparation: under no circumstances should one use the Three Kings Preparation unless one has first used all nine of the biodynamic preparations on the area to be treated. This is a necessity because the nine biodynamic preparations serve to balance as well as to ground and enhance the existing forces, thereby establishing a foundation for the elemental kingdom to build upon. Bear in mind that the biodynamic preparations need to be regarded as forces, not substances.

To apply only the Three Kings Preparation is the equivalent of placing a fence around livestock, thereby securing them from predators, but failing to ensure that the pasture or hay supply is adequate in the area where they are enclosed. That is why it is of utmost importance that one should make an additional commitment to continue to use, in a diligent manner, Steiner's nine basic biodynamic preparations on the area treated with the Three Kings Preparation. When using the Three Kings Preparation, one is in essence sending a message to the elemental kingdom that here, within a "magic circle," the elementals will be provided a safe haven, as well as the profound spiritual nourishment of the biodynamic preparations. Failure to provide such nourishment both before and after using the Three Kings Preparation is tantamount to another betrayal of the elemental world by humanity. *We need not betray them again.*

Preparations as Beings:
The Three Kings Preparation of Hugo Erbe

German-Michael Hahn
Applied Biodynamics, *no. 35, 2001–2002*[1]

The months of harvesting in our time form a gross contrast to those of former times. When, for instance, we take the harvest pictures of Millet, the painter of the peasant life of old, as a message of the intensive human work on the organism of the earth, then the lonesome harvest mood of the modern farmer becomes manifest—he who has to manage the partial and full mechanical procedure, not to mention the consequences and confusion of storage and marketing. If we go back a step further to the previous sowing, we find a corridor, devoid of human beings, that with highly mechanized, fungicide-sprayed seed, should foster growth: reduction of the living organism Earth to, at one's pleasure, an exchangeable substratum for the production of—if necessary, gene-technically altered—FAST FOOD! To fill these "desecrated" corridors with new life, to confront this materialism with idealism, is one of the most excellent strengths of the biological-dynamic preparation-beings in the hands of a spiritual farmer.

When the coming Christmas time with the Twelve Holy Nights approaches its peak, then the festival of the Three Kings comes near. On Three Kings Day, the intense time of inner contemplation can bear its first fruits. Hugo Erbe was the one who cultivated this time many decades ago. He possessed the special gift of contact with the elemental world so that he could introduce an expansion of the preparations created by Rudolf Steiner. In addition to preparations

---

1   Translated from German by Ben Emmett, 1998.

Preparations as Beings: The Three Kings Preparation of Hugo Erbe

that fit directly into the context of the "Agriculture Course" (e.g., calcium, loam, and nitrogen preparation) he also developed preparations that arose from his insights into the elemental world. After the atomic fallout of the 1940s, he noticed suddenly a desertification of his acreage in the Black Forest. He was able at that time to make harmless the demonizing by means of his Three Kings Walk and the substances gold, frankincense, and myrrh.

If now we grind and dynamize the Three Kings' gifts for one hour at midnight of New Year's Eve, we do it trusting a communication of Rudolf Steiner's to Herbert Hahn that tells us that the Folk Spirit frees us for a few minutes during which resolutions made receive a strong force effect toward their realization.[2] The grinding is best done in a large, strong glass or porcelain dish since other mortars are often too small for sufficient force to be generated with the pestle to pulverize the coarse-grained frankincense resin and blend it with the other two substances.

The time from New Year's Day to Three Kings Day belongs to the Twelve Holy Nights and is thereby reserved for spiritual work. Perhaps one or another Demeter (biodynamic) farm can be provided with the dynamized substance in this time period. Thus, further positive centers of power can be created that give life to the countryside.

Then on Three Kings Day, we need to rake the coals in order to warm the stirring water. The following one-hour stirring at the steaming preparation pot belongs to the first peak that the yearly preparation dance introduces.

The freeing of the etheric oils can bring to our attention that the elemental beings are "disenchanted" from the substances and are announced to us. Afterward the storage containers that we bring along are filled and stowed away in the backpack.

After a brief meditation, the participants move along to the open places of the circle, around an area on which the preparations shall

---

2   Herbert Hahn (1890–1970) had contact with Dr. Rudolf Steiner and was among the first teachers of the Waldorf School in Stuttgart.

work. (The circle is divided up into segments, each of which is treated with the preparation by two people.) It is a good idea for each seg-

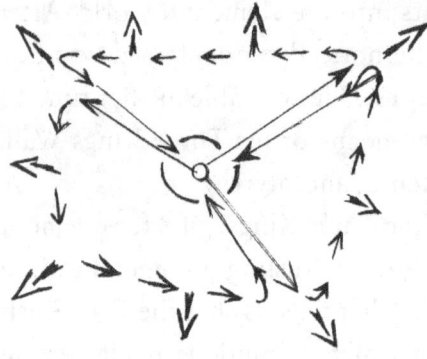

ment to be covered by two people since one person can do the work with the preparation with a small bucket and hand brush, while the other carries the backpack and keeps an eye on the course to be followed. In any case, it would be a good idea, before the actual "walk," to walk around the preparation circle because a consciousness, an experience of the preparation island that is inscribed in the countryside, is found.

In the setting out of the Three Kings Preparation, the liquid is sprayed in the circle toward the outside, so that even in the case of a change of direction of the circle a virtual radiant wreath comes into being. This should have the effect that malevolent elementals lose their strength, and also that they have the possibility of adjusting themselves back into the order.

As the elementals dip into the force field of the radiant wreath, they receive new provisions for going to work on the current problems. In this way we create—as with all other uses of preparations—by free human decision, islands for the working of the elemental world.

Without these possibilities of regeneration, the elemental beings would go hungry, finally turning away from us, turning against us.

Method

- 30 grams Aurum D2
- 30 grams Frankincense
- 30 grams Myrrh

This amount is sufficient for approximately 300 liters of water.

Preparations as Beings: The Three Kings Preparation of Hugo Erbe

- 1 hour dynamizing of gold/frankincense/myrrh at New Year. Afterward, mix with grams of glycerine (derived from organic fat). A viscous emulsion develops that should be sealed in a glass container.
- January 6th—begin stirring from 1:30 p.m. on (because of "elemental rest"; the time the earth is in ethereal breath; verbal communication of Rudolf Steiner to Rully).[3] To liters water we add grams preparation emulsion. In cold weather this will be very thick, so stir well-sunned rain or pond water for one hour (one person), forming a deep vortex and constant change of direction.
- January 6th—2:30 p.m. Fill preparation into containers (plastic gallon bottles are lighter.)
- Distribute preparation into an area of land, approximately 1–2 liters per kilometer. The Three Kings water is sprayed toward the outside of the area every 10 to 50 meters on the vegetation-covered land, also if covered with snow.
- Time needed is about two hours for 5 to 8 kilometers, depending on the kind of terrain.
- The Three Kings water must completely encircle a definite area.
- For the separate segments of the circle there should be exact descriptions prepared, otherwise exact knowledge of the way. At forks and changes of direction clear marking should be set up, on vertical surfaces in case of snow. Return of the various groups should be monitored by one person who knows the whole circuit and in case of stragglers (twisted ankle, for example) can go to meet them.
- For the organizing of parking, it is good if group B drives in the car of group A to point II, then walks to point III. Group

---

3   Franz Rully (Rulniewicz or Rulni) was a coworker in the garden of the Goetheanum and had contact with Dr. Steiner. Later, Franz Rully was a Demeter advisor.

A walks at the same time from I to II and arrives, for example, when it is getting dark. It will be easier for each group to find their own car. Then all go back, return borrowed keys (perhaps with tea and cookies).
- Each person should keep an eye on the whole circuit, with the others, and also in the course of the year.

Protocol for Hugo Erbe's Three Kings Biodynamic Preparation

The protocol for this preparation was sent by German-Michael Hahn, who works intensively with this and the other biodynamic preparations in the Würzburg area of Germany. He manages a 2,000-hectare forest (hectare = 2 ½ acres), with 500 hectares managed biodynamically. He also works with the Sensitive Crystallization method and the Steigbild (capillary dynamolysis) method for quality control.

This preparation is not an "area" preparation (see figure), but rather an "encircling" preparation—that is, it is sprayed *around* the area to protect the boundaries (figure b).

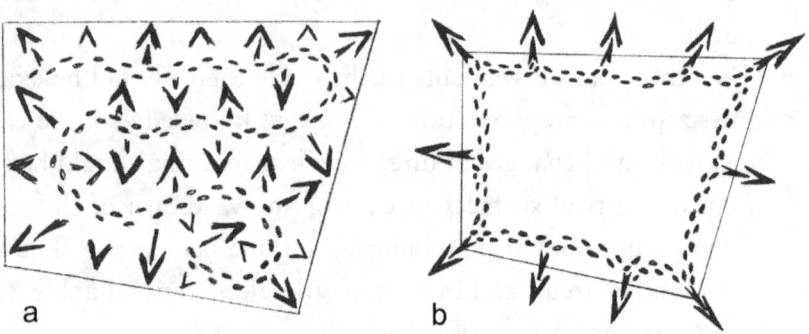

Figure a: An area application of BD 500
Figure b: An encircling, or perimeter, application of the Three Kings Preparation

Preparations as Beings: The Three Kings Preparation of Hugo Erbe

Using a whisk broom, spray the material always toward the outside to keep away adversarial counterforces, and by this means, offer them reconciliation.

The Proportions

    30g Aurum metallicum D2 (gold)
    30g Olibanum (frankincense)
    30g Myrrh (resin myrrh)

In a porcelain mortar, crush the material for one hour. At the end, add:

    50g Rainwater
    50g Plant Glycerine

and continue the process, as above, for the last five minutes.

- The preparation of the gold, frankincense, and myrrh with the mortar and pestle is done on New Year's Eve for one hour from 11:30 to 12:30. This is a very significant time.
- This somewhat fluid substance after a short time becomes a tough paste that can be dissolved only in about 60°C (140°F) water.
- The material should be put in a non-metallic, air-tight container and this will hold the forces; it can be kept for an unlimited time.
- On the 6th of January stir the preparation from 13:00 to 14:00 (11:00 to 13:00 there is an elemental quietude) and then late afternoon, around 16:00, when the etheric in-breathing of the earth organism begins, apply the preparation in the manner described above.

The Amount of Prepared Substance and the Amount of Water

- Use about 5 ccm per liters of water (lukewarm rainwater or water from a pond that has been exposed to sunlight). A further example:

50 ccm use 80 liters H2O
   100 ccm use 160 liters H2O
- The stirring should take place in a non-metallic vessel and the process should be the same as for the stirring of the Bio-Dynamic horn-manure (BD500) and the horn-quartz (BD501) preparations, with the change of direction and strong vortex formation.
- The preparation, in its paste form, must be thoroughly dissolved in 60° C (140° F) water; if it is mixed in the lukewarm water in the larger vessel first, without first being dissolved, it will clump up and not dissolve.
- For 1 km (1,000 m) of boundary 1 to 2 liters of stirred preparation is needed. About:
  - 5 ccm paste covers 4 km of boundary
  - 50 ccm covers 40 km of boundary
  - 100 ccm covers 80 km of boundary
- In covering the perimeter of the territory to be treated, it is best to take a small amount of preparation in a small enough container, so that the amount 1 to 2 liters per 1 km (1,000 m)—is apportioned out.

To Summarize

30 gms.	Gold
30 gms.	Frankincense
30 gms.	Myrrh
50 gms.	Rainwater
50 gms.	Glycerine
190 gms.	Paste

5 ccm for	8 liters H_2O
50 ccm for	80 liters H_2O
100 ccm for	160 liters H_2O

Preparations as Beings: The Three Kings Preparation of Hugo Erbe

Thanks to Jennifer Greene of the Water Research Institute of Blue Hill, Maine, for furnishing and translating the above protocol. Thanks also to Lloyd Nelson for the sketches.

German-Michael Hahn, *like a lawyer, pleas on behalf of trees and forest. He lectures to local councils, calls on mayors, visits farmers and forest owners, and speaks with hunters. The wooded areas he looks after are scattered across 21 towns in the region of Würzburg in Bavaria, an area of 2,500 hectares of communal and private forest, plus he shares responsibility for an equally large area of state forest.*

Expanding the Biodynamic Preparations

Hugh J. Courtney

Applied Biodynamics, *no. 71, winter 2010–2011*

From the very beginning days of biodynamic agriculture, the early pioneers in the movement were concerned with and focused on bringing to Rudolf Steiner's mandate reality:

> The most important thing is to make the benefits of our agricultural preparations available to the largest possible areas over the entire earth, so that the earth may be healed and the nutritive quality of its produce improved in every respect. That should be our first objective. The experiments can come later.[1]

We will take a close look at three of the most prominent of the various biodynamic compound preparations that have been created in different geographic areas of the world by several of the major pioneers of biodynamic agriculture.

In Europe, Maria Thun created a popular biodynamic compound recipe, Barrel Compost (BC), also known by several other names: Cow Pat Pit preparation, *Fladenpreparat*, or simply BC—Biodynamic Compound Preparation (Thun recipe). In Australia, with vast acreages to deal with and a shortage of compostable materials, Alex Podolinsky chose to add the six compost preparations to regular BD500 (horn manure), with astonishingly successful soil improvements to millions of acres, creating deep, humus-rich soils in a time span of

---

[1] As voiced to, and recorded by Ehrenfried Pfeiffer in "We Remember Rudolf Steiner," and subsequently republished in the introduction of *Agriculture,* as translated by George Adams, and in the appendix to *Spiritual Foundations for the Renewal of Agriculture,* translated by K. Creeger and M. Gardner. Excerpt from the G. Adams translation.

years instead of the centuries required by Nature herself. In America, Ehrenfried Pfeiffer, when faced with the difficulty of introducing the actual biodynamic preparations to the practical, "show me" American farmer, formulated the BD Compost Starter and the BD Field Spray. These substances, when introduced, emphasized their bacterial component, and by so doing, created a less esoteric pathway for introducing biodynamic agriculture.

In addition to the inspiration of Rudolf Steiner's original vision, there were other reasons prompting the development of biodynamic compound preparations. In the first place, it was a challenge to produce sufficient biodynamic compost to fulfill the customary biodynamic ideal (as initially practiced in Europe) of first making compost, applying it to the fields, and only then introducing the spray preparations BD500 and, subsequently, BD501 (horn silica). In the second place, it was also necessary to find some way to extend a relatively small quantity of the various compost preparations to meet a large need. A third reason was to improve efficiency of labor as well as to reduce possible soil compaction by making fewer trips over a given field with heavy equipment. Finally, in America, Pfeiffer faced an almost impossible task of trying to explain the biodynamic preparations to the American farmer who found the methods used to make the preparations required a "suspension of disbelief" not easily mustered.

The first publication of Maria Thun's recipe in America appeared in the winter 1981 issue of *Biodynamics* (no. 137), in the article "Compost Preparation According to Maria Thun's Method" by Xavier Florin. A follow up article, "Further Indications Concerning Maria Thun's Compound Preparation," by David Miskell appeared in *Biodynamics* (no. 138), spring 1981.

It was these two articles that inspired me to try my hand at making the Thun Barrel Compost the same year they appeared. The recipe described in the first article consisted of the following ingredients: "five buckets of manure (a bucket of ten liters or approximately ten quarts), five hundred grams (= 1.1025 Av. Pds.) of basalt powder, and one hundred grams (= 0.2205 Av. Pds.) of powder of eggshells."

Initially, I could not find a source of basalt powder, and so used local quarry granite dust in making my first ever batch of barrel compost. Subsequently, with the help of John Carlson, I was able to locate sources of basalt powder in both Minnesota and Massachusetts. Although the first several batches of BC produced here were made using wooden whiskey or wine barrels, it became apparent that one could use the same barrel only about three times before it became necessary to replace it because the action of the BC in the barrel resulted in composting the interior of the barrel quite successfully. (According to Maria Thun's experience, the bottom of the barrel must also be removed, so that the manure is in contact with the soil; otherwise, transformation does not take place.) After experiencing such rapid deterioration of the barrels, I chose to line the pit with bricks.

> Basalt is a rock of volcanic origin and sources can be determined by consulting your state geologist to see what road gravel quarries within the state may be quarrying from a basalt rock site. Should a quarry identify itself as having "trap" rock, it is likely of volcanic origin and basaltic in nature. Every quarry in each state is required to keep on file an analysis of all of its constituents that compose 1% or more of content.

In India—no doubt for similar reasons and probable termite damage—Peter Proctor used brick to line numerous contiguous pits when making the "Cow Pat Pit" (CPP) preparation.[2]

Our current practice is to make a circular pit lined with brick, using a total of eighty-one bricks per pit, sunk into the ground to a depth of four and a half bricks up to ground level with four and a half bricks above the initial soil level and banked with the soil removed from the pit. The circle of bricks uses nine bricks

---

[2] As can be seen in the movie, *How to Save the World*.

Expanding the Biodynamic Preparations

Left: Girl Scouts stirring manure for BC
Right: Adding compost preparations to BC

for each level, and a total of nine levels of brick are needed from lowest to highest tier. Into this pit, we place two full recipes, consisting of twelve and a half gallons of manure per recipe for a total of twenty-five gallons of manure per pit. This manure has been stirred for one full hour, incorporating one thousand grams of basalt powder and two hundred grams of ground eggshells. To this manure mixture, we add two full sets of the compost preparations, although the BD507 (valerian) we now put in the mixture only at the halfway or forty-five-day point when the contents have assumed a more colloidal humus-like nature. Also, at that halfway point of working with the BC mixture, we will add an additional set of compost preparations. Although some practitioners favor the inclusion of stinging nettle herb added to the manure along with the basalt and eggshells, practical application on actual crops using this form of a compound preparation leads to an increased emphasis in the growth of the leaf and stem portion of the plant, a result that, while favorable for some crops, is definitely not desired in vineyards and perhaps other fruiting crops.

In the late 1980s, when I was privileged to meet Maria Thun in Kimberton, Pennsylvania, I was advised to extract the BC from the pit while it still had substantial moisture, and a very clay-like consistency, and to use egg shells not ground to an extremely fine powder, but to maintain a particle size that could still be clearly identified as belonging to an eggshell. Such a particle size would be about the same size circumference as the lead in a pencil. (Thun credits discussions with Pfeiffer as having led to the inclusion in her recipe of eggshells as a form of living calcium.) I was also advised by her to set the price of a unit of the BC at the same level as charged for a unit of BD500. Maria Thun has claimed from the beginning that the pre-stirring of the manure, basalt powder and eggshells allows one to apply the finished BC after a mere twenty minutes of BD stirring. Regrettably, she has not provided any data or other substantiation to support that conclusion on her part. She is also adamant that the BC should not be stirred together with BD500. Many practitioners, however, choose to stir BC and BD500 together for one full hour, or more often than not, the preference is to add the BC for the last twenty minutes when stirring the BD500. My own conclusion is that the first time one is treating soils, the two different preparations should preferably not be stirred together, but subsequent applications could be stirred together. Inasmuch as the energies of BC and BD500 are already working together in the soils, a combined application should not have the same possibility of interfering with each other.

Reducing the number of times heavy equipment might contribute to soil compaction is a significant factor to be considered. For many years now, the BC has been serving as an effective way to introduce the energy of the biodynamic compost preparations to the soil, and we have consistently recommended it, especially for those beginning to practice biodynamic agriculture.

On another continent, Australia, Alex Podolinsky chose a different approach to treat the soil with the energies of the six compost preparations as well as the BD500. In thirty-gallon copper containers full of the finished BD500, he added six sets of the BD502–507. Since the finished BD500 has already arrived at a condition of colloidal humus, there is no need to hold the BD507 in reserve. Although there are those who object to his formulation, a careful reading of the Agriculture Course would seem to provide substantial justification for Podolinsky's approach in Steiner's own words:

> But in other ways, too, we must still try to give the manure the right living property. We must give it such a consistency that it will retain of its own accord as much of nitrogen and other substances as it requires. For we shall thereby impart to the manure a tendency to that living vitality that will enable it to bring the right vitality into the Earth itself. Today therefore—more as a general indication—I shall mention a few more things in the same direction: preparations to add to the manure itself in minute doses, in addition to the cow-horn stuff. The preparations we add to the manure vitalize it in such a way that it will be able to transmit its vitality to the soil from which the plants are springing.[3]

While it is easy to conclude from the foregoing that Steiner is referring to the process of dealing with barnyard manure to be spread out on the fields in the customary fashion of "manuring" (or fertilizing) as practiced at the time, there is also nothing he says that would not also be applicable to the manure that comes from the recently buried cow horns as well. Given the vast acreages of sheep and cattle ranches in Australia that make it virtually impossible to apply sufficient BD-made compost, the Podolinsky solution seems to be highly appropriate. After all, what one is after is the distribution of the forces of the biodynamic preparations, not the substances thereof.

However, certain objections have arisen to Podolinsky's "Prepared" BD500 (500P) in some European circles. As reported by Monty Waldin in his book, *Biodynamic Wines*:

---

3   Steiner, *Agriculture Course*, p. 91.

Opponents of the prepared cow manure, like Bouchet, say it confuses the soil by creating a growth-decomposition opposition between horn manure, which stimulates growth of the roots in the soil, and the Biodynamic compost preparations, which stimulate decomposition of compost material. This causes the plant roots to remain in the topsoil. This may suit growers of annual crops like vegetables, but is not ideal for perennial crops like vines reliant on a permanent and deep root system. Bouchet also points out that Steiner outlined how both horn manure and the Biodynamic compost preparations should be made, but he never said anything about prepared horn manure.

I could point out that there are many things that Steiner never explicitly said in the lectures of the *Agriculture Course*, but may be inferred by exercising careful thought and taking into account what he may have said in some of the other thousands of lectures he gave. In addition to Bouchet's opposition, a Dr. Christian v. Wistinghausen of Mausdorf, Germany sent out a widely distributed commentary entitled "500 P: A Mistake?" This commentary states that the motivation behind such a combination is merely to expedite Demeter recognition, and further, it uses essentially the same wording as is quoted above from Bouchet in condemning 500 P. To quote from the commentary:

> Horn Manure promotes root and plant growth and their life force, supports organisms in the soil that encourage plant growth and creates humus and the "I" potentiality in the plants cultivated.... The compost preparations have the task of building up the useful humus by organizing cosmic forces. They support organisms in the soil, which break down all organic substances and, together with mineral clay substances, provide for creation of the clay humus complex in the soil.... Horn manure and compost preparations result in opposing processes. If the processes are mixed during their creation, they will cancel each other out.... The use of "500 P" or "P 500" does not adhere to Demeter guidelines, since it is no longer horn manure.

Expanding the Biodynamic Preparations

There is no evidence that Dr. Wistinghausen ever used 500 P before reaching the above conclusions, and one wonders whether such conclusions are mere "abstract intellectualism," especially since no foundation has been laid to substantiate that the processes "will cancel each other out."

As far as Bouchet's comment that use of the 500 P "causes the plant roots to remain in the topsoil," given the totally opposite experience of creating humus-filled soils and very deep-rooted plants, not only in Australia, but here in the United States as well, one can only question the quality of the 500 P and the compost preparations used by Bouchet to arrive at such a conclusion, assuming that 500 P was actually used by him. Alex Podolinsky, himself, brings into question the quality of some of the biodynamic preparations being made in Europe.[4]

It is of further interest to note that Dr. Wistinghausen's paper on 500 P is accompanied by an additional sheet that promotes the use of a product labeled "Mausdorf Activator." One wonders again whether the motivations for condemning 500 P are not dictated by a modicum of self-interest in promoting a competing product.

Sometime in 2007, it came to my attention that a vote had been taken within a Demeter organization in Europe not to allow Demeter certification where "Prepared" 500 had been in use (although the product is allowable under Demeter USA standards). In response to that information, I wrote in protest to Nikolai Fuchs, the head of the Agriculture Section of the School of Spiritual Science of the Anthroposophical Society in Dornach, Switzerland, as follows:

> I am aware of a recent vote that does not endorse the use of Alex Podolinsky's Prepared 500. I observe that this appeared to be an exclusively European vote, with no input from America or Australia. How unfortunate that this matter was not put to a practical test, but relied on mere human opinion from persons who in all likelihood had never used the product. I would respectfully

---

4   Podolinsky, *Bio-Dynamics: Agriculture of the Future*, p. 18.

suggest that the matter be reopened and that a real examination be used to make such a decision rather than relying on opinion not based on direct perception. Prepared 500 has been used by at least a few people in this country with excellent results, and the formula itself can certainly be justified by a careful reading of Rudolf Steiner's own words. Again, I would urge reexamination of the question.

To date, I have never received any concrete response from anyone in Europe that my suggestion was taken up, and insofar as known, 500P is still not allowed for Demeter certification in Europe, even though there are now a number of farms and especially vineyards in Europe that are using the Prepared 500 made in accordance with Alex Podolinsky's standards, with the same humus building and deep-rooted plants resulting.

One last point of information on the Prepared 500 needs to be mentioned. It was in 1989, I believe, when Podolinsky shared with me his recipe for Prepared 500 at a conference in Kimberton, where he was lecturing. At that time, his practice was to use nothing but regular BD500 on a farm's soils for the first two years, with a switch to Prepared 500 thereafter. In a recent telephone conversation, Alex advised me that further experience in using the Prepared 500 had established that one could begin to apply it even in the first year of transitioning to biodynamic practice. The only exception to this change in practice was reserved for soils that were virtually sterile, or so poisoned and maltreated that an application of regular BD500 alone was still necessary to restore some semblance of life to those soils. Certainly, the results achieved by many farms and vineyards in Australia and many other parts of the world that have taken up the use of Prepared 500 provide overwhelming substantiation for the benefits to be obtained from using it in one's biodynamic efforts.

Our last look at biodynamic compound preparations involves an examination of Ehrenfried Pfeiffer's contribution to this subject.

Expanding the Biodynamic Preparations

Pfeiffer was a highly qualified scientist and bacteriologist, and made significant contributions to several fields of science, most especially filter paper chromatography and sensitive crystallization techniques for analysis. Sometime in the late 1940s, because of the disinterest that greeted efforts to promote the use of the biodynamic preparations, Pfeiffer began the development of a compost starter material that incorporated the compost preparations and BD500 in a more easily used and more stable form than was possible using the actual preparations. In the process of developing what subsequently became known as the BD Compost Starter, he extracted, from the preparations themselves, a certain number of strains of bacteria that proved quite beneficial in creating good compost that had a very high proportion of colloidal humus. Ultimately, he extracted and cultured some fifty-five different bacteria and other organisms, virtually all of which were found in the compost preparations and the BD500. These organisms continue to be cultured and are still used in the production of the Pfeiffer BD Compost Starter.

The Starter also includes a significant dosage of the actual preparations in its manufacture. As a companion or corollary to the Compost Starter, Pfeiffer also formulated the Pfeiffer BD Field Spray for use on farm fields. It proved to be especially useful when turning under green manure crops. In its production, the Field Spray contained a much higher proportion of the BD500 and a much-reduced proportion of the compost preparations BD502–507 than was the case for the BD Compost Starter. When Pfeiffer originated these formulas, it was much easier to promote them to the American farmer for the "beneficial bacteria and enzymes" they contained rather than making reference to the biodynamic preparations.

Along the way, there have been several attempts at "industrial espionage" by individuals who sought to obtain the secret behind the efficacy of the Pfeiffer products. In every case, the focus was on the "bacteria" and not on the biodynamic preparations they contained. And, in every case, an inferior product resulted from these espionage efforts, because the individuals chose to "keep the bath water" and "throw away the baby."

In the same fashion as one stirs or potentizes the manure, eggshells, and basalt powder in making the Maria Thun Barrel Compost, so do the Pfeiffer products also go through a much more rigorous potentization process when they are made. Further justification for Pfeiffer's efforts to produce the BD Compost Starter and BD Field Spray can be found in the minutes of a July 1955 Board of Directors meeting of the Biodynamic Farming and Gardening Association:

> [Dr. Pfeiffer] worked out 27 different tests to determine quality on the BD preparations. These are mostly to test biological value, only the crystallization test being used to determine dynamic efficacy. He has come to the conclusion as a result of this recent work that only experts should make the preparations, not every farmer his own.

Given the fact that by 1955 there were undoubtedly some amateurish and poor quality preparations being made, it is probable that Dr. Pfeiffer felt a need to establish a fail-safe baseline of quality to achieve an acceptable and consistent biodynamic result. In his own way, he was trying to meet Rudolf Steiner's mandate "to make the benefits of our agricultural preparations available to the largest possible areas over the entire earth."

In 1995, when the Pfeiffer Foundation was being evicted by Sunbridge College (with the help of the local building inspection department) from the ramshackle conglomerate of buildings that housed them in Spring Valley (now Chestnut Ridge), New York, JPI and the Courtney farm offered to provide a place to house the continued production of the Pfeiffer materials. Subsequently, an agreement was reached with the Pfeiffer Foundation for JPI to distribute and produce batches of the BD Compost Starter and the BD Field Spray. Thereafter, the Pfeiffer Foundation ceased all operations in Spring Valley. Not long after this agreement was reached, JPI began to experience

Expanding the Biodynamic Preparations

very strong expressions of antipathy to the continuation of Pfeiffer formula production. Comments were heard from several unexpected quarters along the lines of "Why don't you just let it die?" It was during this period, as well, that we learned that there was a substantial group of people who regarded Pfeiffer as a person who had betrayed Steiner's work by putting the BD Compost Starter and BD Field Spray out in the market place and thereby "commercializing" biodynamics.

Given the virtual poverty that Pfeiffer experienced in his later years and his exceptional dedication to the biodynamic work, it is difficult to agree with such a characterization. Instead of viewing the joining together of Josephine Porter's preparation work and Pfeiffer's BD Compost Starter and BD Field Spray work as an extremely important event with not only physical world consequences, but highly significant from a spiritual point of view, JPI has continued to be maligned in certain quarters as having also betrayed the biodynamic work. This attitude is what it is, and JPI will continue to provide a "home for the preparations" as was Josephine Porter's hope. JPI will also be honored to provide a home for Pfeiffer's contribution of the BD Compost Starter and BD Field Spray.

While the agreement with the Pfeiffer Foundation clearly specified the proprietary nature of the Pfeiffer products, JPI has continued the production of these materials in the belief that they provide an easily opened door for mainstream agriculture to take up biodynamic practices. In more than a few instances, using these products has served to stimulate a greater interest in how they achieve such effects, and there have been several farmers who have gone beyond their use and begun to take up the use of the actual biodynamic preparations themselves. During 2010, JPI established clear legal ownership of the Pfeiffer formula, and is now in a position, if deemed desirable, to license its production to other entities. In 2010, JPI also chose to change the name of the Pfeiffer BD Field Spray to the Pfeiffer BD Field and Garden Spray, because of its widespread and successful use in gardening applications.

As one of the premier biodynamic compound preparations available in the world, Pfeiffer's contributions of the BD Compost Starter and the BD Field and Garden Spray deserve a place in the continuing expansion of biodynamic horizons along with Maria Thun's Barrel Compost and Alex Podolinsky's Prepared 500. All three of these biodynamic pioneers have made a significant contribution to biodynamic agriculture, and each of them has done an admirable job of fulfilling Rudolf Steiner's hope that the biodynamic preparations would serve humankind "so that the Earth may be healed."

Amethyst 501

Dennis Klocek
Applied Biodynamics, *no. 12, 1995*

In the summer of 1993, an experiment was undertaken to see if the potentization of the gem amethyst would have an effect upon plant growth in the climatic niche of Central Valley, California.

The goal was to see if a variation of the BD501 spray could be made that would provide a more moist light under the arid intense light conditions of a central California summer.

The crop chosen was cabbage. It was sown in May, a usually foolhardy proposition in that the growing temperatures are in the 100s or high 90s from early July until mid-October; that is, during the cabbage maturation period. Normal cabbage sowing is done in mid-July and the plants persevere in the heat until cooler fall temperatures provide the dew and coolness that they require.

The amethyst was chosen because of its cool shadowy violet color. It is a gemstone in the silicate family that has an interesting chemical profile that suggested it would be a good spray for plants, especially cabbages. The chemical/mineral constituents of amethyst are: 1. silica; 2. lime; and 3. soda. The metals are: 1. iron peroxide; and 2. manganese. Silica is present in the biodynamic preparation BD501 as the carrier of light forces into the plant. In like manner, amethyst is a silicate. It differs from rock crystal in having a calcium radiation in its chemistry. This calcium radiation is a primary gesture. In amethyst, the silica is united to calcium. The amethyst also has iron peroxide in its structure. Iron, especially in the form of peroxide, has an especially strong relationship to the element oxygen or "life" in Rudolf Steiner's alchemy found in *Agriculture Course*.

Currently, there are a number of peroxide "cures" and soil treatments for plants and animals in the "fringe" literature. The basic idea behind using peroxide as a soil drench for animal food is to increase the ability of oxygen to become available to organisms in order to improve their respiration. In amethyst, this extra oxygen is wedded to iron, the breathing metal.

Plants like cabbages, which make heavy demands on calcium in the soil, often also make equally heavy potassium demands for midribs of leaves and strong root-stem development. Soda or sodium acts in such a way that it helps to change potassium from an insoluble form to a soluble form in the soil. As a mineral catalyst then, amethyst is a silicate with a calcium/potassium gesture in its chemistry. The metals iron peroxide and manganese both show an affinity for enhancing life or vitality, one of the old uses for amethyst jewelry.

The manganese, which is the actual source of the deep violet hue of the gem, has the added property of being particularly suited for relating the sulfuric acid and the oxidizing forces in metallurgy. According to Walther Kloos in *The Living Earth*, manganese in nature is always found in combination with oxygen and represents a gesture among metals that indicates support for the vegetative process.

By contrast, metals with an affinity for sulfur show affinities for the general (peripheral) life forces of flower/fruit and seed. This again puts the amethyst into close relationship to the life gesture of the cabbage, which is to produce leaf after vegetative leaf before it rises on its stalk into flowering.

By simply comparing the growth gesture between quartz crystal and amethyst, this contrast can be observed.

The quartz breaks out into open stamen-like shafts of light as it grows, indicative of a strong seed/light force. By contrast, the amethyst is most often found in geodes pointing inward toward the center. Its crystals, while they show beautiful points, rarely grow into discrete shafts but instead cluster together in planes of points without shooting into space. This is an analogy for the ideal cabbage gesture, a holding of form in one plane before shooting up vertically.

From this, it can be seen why the amethyst appeared to be a good cabbage spray.

The experiment sought to retain and possibly amplify these qualities into the grinding process used to render the substance into gem flour. Three nice amethyst points were pulverized and ground between glass plates and then worked in a mortar and pestle for one hour. A very fine, colorless meal resulted. One tablespoon of this was saved and to this, as a dilutant, 10 tablespoons of orthoclase feldspar were added. The orthoclase is a potassium silicate with traces of calcium. This was used to enhance the potash/calcium action of the spray. This was triturated in a mortar and pestle for one hour. (*Triturate: to rub, crush, or grind into a very fine powder/pulverize.*)

One tablespoon of this was taken and 10 tablespoons of flint were added as a dilutant and one more hour of trituration brought the amethyst to a 3x potentization.

This meal was put into a cow horn and buried for the summer. Before burial, some of the 3x flour was taken to spray on the cabbage crop. This amethyst spray was applied on leaf days (when the moon is in a water sign) after 20 minutes of stirring in rainwater.

Due to extreme dryness, the seedling cabbages were often sprayed in the evening. In the morning strong dew would always be present in the cabbage patch. The cabbages were mulched heavily and watered by a drip system under the mulch. They were side-dressed with a liquid ferment of alfalfa/kelp/and rotted cabbages that had stabilized completely before applying it to the soil around the plants.

The cabbages grew steadily through a very hot (the high 90s/low 100s) summer without missing a beat. They produced enormous heads by September 1. The savoy heads were sweet, crisp, and well filled out. The only blemish was that in some of them the sun had scalded some leaves when the head was forming and those ended up in the cabbage. An overhead shadow cloth would have corrected this flaw. The sweetness and perfection of growth belied their hot arid growing period. When they were cut open, the open savoy spaces were filled with shining beads of water, even though no water had touched the leaves or heads during the whole growing period except for dew-fall.

Amethyst sprayings were alternated with Barrel Compound in the spraying cycles. Ferments were applied during the waxing phases of the moon until head formation was well underway. No doubt, clay loam with a good heart, heavy mulching and drip irrigation were positive factors in the growth of the cabbages. However, the extreme climatic conditions that they endured prompted very experienced local gardeners to label the idea "doubtful." However, the size, sweetness, and perfection of a bed full of savoy salad cabbages in early October in our part of California pointed to some unusual gardening techniques.

Next spring the first "horn amethyst" will be used on an over-wintering savoy crop that is now sleeping in the rain. And, next summer, horn amethyst will once again be called upon to produce positive results from a doubtful experiment.

Dennis Klocek *is a teacher, researcher, artist, gardener, and alchemist. He graduated with an MFA in 1975 from Temple University's Tyler School of Art. He then taught for seven years at a community college. In 1982, his love for the work of Rudolf Steiner took him to Rudolf Steiner College in Sacramento, California, where he has been the director of their Consciousness Studies Program ("Goethean Studies") since 1992. Dennis is engaged in research, teaching, and writing in many fields, including weather, gardening, meditation, the human organism, and self-transformation. He founded the Coros Institute to teach and promote dialogue experiences based on esoteric wisdom. Dennis Klocek's many books include* Sacred Agriculture: The Alchemy of Biodynamics *(2012) and* Climate: Soul of the Earth *(2010).*

Cited Works

Barfield, Owen. *Saving the Appearances: A Study in Idolatry*. New York: Faber and Faber, 1957.

Castelliz, Katherine. *Life to the Land*. Sussex, UK: Lanthorn Press, 1980.

Coats, Callum. *Living Energies: Viktor Scahuberger's Brilliant Work with Natural Energy Explained*. Dublin: Gill Books, 2001.

Cocannour, Joseph A. *Weeds: Guardians of the Soil*. Midwest Journal Press (selfhelpbook.midwestjournalpress.com), 2015.

Corrin, George. *Handbook on Composting and the Bio-Dynamic Preparations*. London: Bio-Dynamic Agricultural Association, 1960.

Dodoens, Rembert. *A New Herbal, or Historie of Plantes: Wherein is Contained the Whole Discourse and Perfect Description of All Sorts of Herbes...* (trans. Henrie Lyte), London, 1554.

Duke, James A. *CRC Handbook of Medicinal Herbs*. Boca Raton: CRC Press, 1985.

Erbe, Hugo. *Präparate zur Förderung des elementaren Kräftewirkens im biologisch-dynamischen Land- und Gartenbau*. Tellingstedt, Germany: Lohengrin-Verlag, 2003.

Gail, Peter A. *The Dandelion Celebration: A Guide to Unexpected Cuisine*. Kingsville, OH: Goosefoot Acres, 1994.

Gibbons, Euell. *Stalking the Wild Asparagus*. Guilford, CN: Stackpole, 2020.

Goethe, Johann Wolfgang von. *The Metamorphosis of Plants*. Cambridge, MA: MIT, 2009.

Keyserlingk, Adalbert von. *The Birth of a New Agriculture: Koberwitz 1924 and the Introduction of Biodynamics*. Forest Row, UK: Temple Lodge, 2009.

———. *Developing Biodynamic Agriculture: Reflections on Early Research*. Forest Row, UK: Temple Lodge, 1999.

Koepf, Herbert H. *Bio-Dynamic Sprays*. Kimberton, PA: Bio-Dynamic Farming and Gardening Association, 1971. Currently contained in *Koepf's Practical Biodynamics: Soil, Compost, Sprays, and Food Quality*. Edinburgh: Floris Books, 2012.

Cited Works

———. *The Biodynamic Farm: Agriculture in Service of the Earth and Humanity.* Hudson, NY: Anthroposophic Press, 1994.

Kolisko, Lily. *Agriculture of Tomorrow.* Bournemouth, UK: Kolisko Archive, 1978.

König, Karl. *Earth and Man.* Wyoming, RI: Bio-Dynamic Literature, 1982

Lachman, Gary. *Caretakers of the Cosmos: Living Responsibly in an Unfinished World.* Edinburgh: Floris Books, 2013.

Lievegoed, C. B. J. "The Working of the Planets and the Life Processes in Man and Earth" (pamphlet). Experimental Circle of Anthroposophical Farmers and Gardeners, 1951.

Mowrey, Daniel B. *The Scientific Validation of Herbal Medicine.* Toronto: Cormorant, 1986.

Pasquini, Chris. *Atlas of Bovine Anatomy.* Eureka, CA: Sudz, 1982.

Pfeiffer, Ehrenfried. *Biodynamic Farming and Gardening: Renewal and Preservation of Soil Fertility,* 4th ed. Hudson, NY: Portal Books, 2021.

———. *Pfeiffer's Introduction to Biodynamics.* Edinburgh: Floris Books, 2011.

———. *Soil Fertility, Renewal and Preservation: Bio-Dynamic Farming and Gardening,* 2nd ed. Lanthorn Press

———. *Weeds and What They Tell Us.* Biodynamic Farming and Gardening Association. 1970; currently published by Floris Books, 2012.

Podolinsky, Alex. *Bio-Dynamic Agriculture Introductory Lectures* (3 vols.). Sydney, Australia: Gavemer Foundation, 1985.

———. *Living Knowledge.* Powelltown, Australia: Bio-Dynamic Agricultural Association of Australia, 2002

Prescott, Frederick. *Modern Chemistry.* London: Sampson Low, Marston, 1932.

Reader's Digest Practical Guide to Home Landscaping. Pleasantville, NY: Reader's Digest, 1989.

Rittelmeyer, Friedrich. *Rudolf Steiner Enters My Life.* Edinburgh: Floris Books, 2013.

Sanchez, Anita. *The Teeth of the Lion: The Story of the Beloved and Despised Dandelion.* McDonald and Woodward, 2006

Sattler, Friedrich, and Eckhard Wistinghausen. *Bio-Dynamic Farming Practice.* Stuttgart: Bio-Dynamic Agricultural Association, 1989.

Scott, Timothy Lee. *Invasive Plant Medicine: The Ecological Benefits and Healing Abilities of Invasives.* Rochester, VT: Healing Arts Press, 2010.

Selg, Peter. *The Agriculture Course, Koberwitz, Whitsun 1924: Rudolf Steiner and the Beginnings of Biodynamics*. Forest Row, UK: Temple Lodge, 2010.

Steiner, Rudolf. *Agriculture: Spiritual Foundations for the Renewal of Agriculture* (CW 327). Trans. Catherine E. Creeger and Malcolm Gardner. Kimberton, PA: Biodynamic Farming and Gardening Association, 1993.

———. *Agriculture Course: The Birth of the Biodynamic Method* (CW 327). Trans. George Adams. Forest Row, UK: Rudolf Steiner Press, 2012.

———. *The Archangel Michael: His Mission and Ours*. Ed. Christopher Bamford. Hudson, NY: Anthroposophic Press, 1994.

———. *The Art of Lecturing* (CW 339). Trans. Maria St. Goar and Peter Stebbing. Spring Valley, NY: Mercury Press, 1994.

———. *The Case for Anthroposophy: Extracts from "Riddles of the Soul"* (CW 21). Trans. Owen Barfield. Great Barrington, MA: Chadwick Library Edition, 2018.

———. *Goethe's World View* (CW 6). Trans. William Lindeman. Spring Valley, NY: Mercury Press, 1994.

———. *Harmony of the Creative Word: The Human Being and the Elemental, Animal, Plant, and Mineral Kingdoms* (CW 230). Trans. Matthew Barton. Forest Row, UK: Rudolf Steiner Press, 2002.

———. *How to Know Higher Worlds: A Modern Path of Initiation* (CW 10). Trans. Christopher Bamford. Hudson, NY: Anthroposophic Press, 1994.

———. *Introducing Anthroposophical Medicine* (CW 312). Trans. Catherine E. Creeger. Great Barrington, MA: SteinerBooks, 2010.

———. *Intuitive Thinking as a Spiritual Path: A Philosophy of Freedom* (CW 4). Trans. Michael Lipson. Hudson, NY: Anthroposophic Press, 1995.

———. *Man as Symphony of the Creative Word* (CW 230). Sussex: Rudolf Steiner Press, 1991; currently published as *Harmony of the Creative Word: The Human Being and the Elemental, Animal, Plant, and Mineral Kingdoms*. Forest Row, UK: Rudolf Steiner Press, 2002.

———. *Nature's Open Secret: Introductions to Goethe's Scientific Writings* (CW 1). Trans. John Barnes and Mado Spiegler. Great Barrington, MA: SteinerBooks, 2010.

———. *An Outline of Esoteric Science* (CW 13). Trans. Catherine E. Creeger. Hudson, NY: Anthroposophic Press, 1997.

———. *The Philosophy of Freedom: The Basis for a Modern World Conception* (CW 4). Trans. Michael Wilson. Forest Row, UK: Rudolf Steiner Press, 2011.

Cited Works

———. *Secret Brotherhoods and the Mystery of the Human Double* (CW 178). Forest Row, UK: Rudolf Steiner Press, 2004.

———. *Spiritual Science and Medicine* (CW 312). London: Rudolf Steiner Press, 1948; currently published as *Introducing Anthroposophical Medicine*. Great Barrington, MA: SteinerBooks, 2010.).

———. *The Threefold Social Order* (CW 23). Trans. Frederick Heckel. Spring Valley, NY: Anthroposophic Press, 1972.

———. *Theosophy: An Introduction to the Spiritual Processes in Human Life and in the Cosmos* (CW 9). Trans. Catherine E. Creeger. Hudson, NY: Anthroposophic Press, 1994.

Schumacher, E. F. *Small is Beautiful: Economics as if People Mattered*. New York: Harper Perennial, 2010.

Thun, Titia, and Matthias Thun. *The Maria Thun Biodynamic Almanac: North American Edition*. Edinburgh: Floris Books, annual publication.

Thun, Maria. *The Biodynamic Sowing and Planting Calendar: Working with the Stars*. Bio-Dynamic Association, annual publication.

———. *The Biodynamic Year: Increasing Yield, Quality and Flavour: 100 Helpful Tips for the Gardener or Smallholder*. Forest Row, UK: Temple Lodge, 2010.

———. *Gardening for Life: The Biodynamic Way*. Stroud, UK: Hawthorn Press, 2000.

Wistinghausen, W C. von, et al. *The Biodynamic Spray and Compost Preparations: Directions for Use*. UK: Bio-Dynamic Agricultural Association, 2003.

Further Reading

Berg, Peter. *The Moon Gardener: A Biodynamic Guide to Getting the Best from Your Garden.* Forest Row, UK: Temple Lodge, 2012.

Berrevoets, Erik. *Wisdom of the Bees: Principles for Biodynamic Beekeeping.* Great Barrington, MA: SteinerBooks, 2009.

Bloom, John. *Saucy Tomatoes and Blueberry Thrills: A Humorous Harvest from the Biodynamic Farm.* Great Barrington, MA: SteinerBooks, 2014.

Bresette-Mills, Jack. *Sensitive Beekeeping: Practicing Vulnerability and Nonviolence with Your Backyard Beehive.* Great Barrington, MA: Lindisfarne Books, 2016.

Code, Jonathan Michael. *Muck and Mind: Encountering Biodynamic Agriculture: An Alchemical Journey.* Great Barrington, MA: Lindisfarne Books, 2014.

Damon, Betsy. *Water Talks: Empowering Communities to Know, Restore, and Preserve their Waters.* Spencertown, NY: Portal Books, 2014.

Florin, Jean-Michel. *Biodynamic Wine Growing: Understanding the Vine and Its Rhythms.* Edinburgh: Floris Books, 2021.

Frazier, Louise. *Louise's Leaves: A Cook's Journal around the Calendar with Local Garden Vegetable Produce.* Kimberton, PA: Biodynamic Farming & Gardening Association 1995.

Hauk, Gunther. *Toward Saving the Honeybee,* 2nd ed. Kimberton, PA: Biodynamic Farming & Gardening Association, 2017.

Hoffmann, Nigel. *Goethe's Science of Living Form: The Artistic Stages.* Ghent, NY: Adonis Press, 2013.

Holdrege, Craig. *Thinking Like a Plant: A Living Science for Life.* Great Barrington, MA: Lindisfarne Books, 2013.

Hurter, Ueli. *Biodynamic Preparations around the World: Insightful Case Studies from Six Continents.* Edinburgh: Floris Books, 2021.

Karlsson, Britt, and Per Karlsson. *Biodynamic, Organic and Natural Winemaking: Sustainable Viticulture and Viniculture.* Edinburgh: Floris Books, 2014.

Klett, Manfred. *The Foundations and Principles of Biodynamic Preparations.* Edinburgh: Floris Books, 2023.

Further Reading

———. *Principles of Biodynamic Spray and Compost Preparations*. Edinburgh: Floris Books, 2005.

Klocek, Dennis. *Climate: Soul of the Earth*. Great Barrington, MA: Lindisfarne Books, 2011.

———. *Sacred Agriculture: The Alchemy of Biodynamics*. Great Barrington, MA: Lindisfarne Books, 2013.

Koepf. Herbert H. *The Biodynamic Farm: Agriculture in Service of the Earth and Humanity*. Hudson, NY: Anthroposophic Press, 1994.

———. *Companion Plants: An A to Z for Gardeners and Farmers*. Edinburgh: Floris Books, 2016.

König, Karl. *Social Farming: Healing Humanity and the Earth*. Edinburgh: Floris Books, 2014.

Kranich, Ernst Michael. *Planetary Influences upon Plants: A Cosmological Botany*. Kimberton, PA: Biodynamic Farming & Gardening Association, 1986.

Kuepper, George. *Psychotronics and a Biodynamic Garden: How to Grow and Harvest Healthier Food through Radionics and Dowsing*. Great Barrington, MA: Portal Books, 2014.

Lorenzen, Iwer Thor. *The Spiritual Foundations of Beekeeping*. Forest Row, UK: Temple Lodge 2017.

Mansvelt, Jan Diek van. *Wonders of Development in Plants, People, and Projects*. Ghent, NY: Adonis Press, 2022.

Massei, Karsten. *Gifts of the Honeybees: Their Connection to Cosmos, Earth, and Humankind*. Spencertown, NY: SteinerBooks, 2022.

Masson, Pierre. *A Biodynamic Manual: Practical Instructions for Farmers and Gardeners*. Edinburgh: Floris Books, 2014.

Mizon, Bob. *Stargazers' Almanac 2024: A Monthly Guide to the Stars and Planets*. Edinburgh: Floris Books, annual publication.

Moora, Walter. *A Farmer's Love: Living Biodynamics and the Meaning of Community*. Great Barrington, MA: Portal Books, 2011.

Morrow, Joel. *Vegetable Gardening for Organic and Biodynamic Growers: Home and Market Gardeners*. Great Barrington, MA: Lindisfarne Books, 2014.

O'Connell, Marina. *Designing Regenerative Food Systems: And Why We Need Them Now*. Stroud, UK: Hawthorn Press 2022.

Pfeiffer, Ehrenfried E. *Bio-Dynamic Gardening and Farming* (2 vols.). Spring Valley, NY: Mercury Press, 1994, 1995.

———. *Chromatography Applied to Quality Testing.* Kimberton, PA: Biodynamic Farming & Gardening Association, 1984.

———. *Sensitive Crystallization Processes: A Demonstration of Formative Forces in the Blood.* Spring Valley, NY: Anthroposophic Press, 1975.

Pfeiffer, Ehrenfried E., and Michael Maltas. *The Biodynamic Orchard Book.* Edinburgh: Floris Books, 2013.

Philbrick, John, and Helen Philbrick. *Gardening for Health and Nutrition: An Introduction to the Method of Biodynamic Gardening.* Hudson, NY: Anthroposophic Press, 1995.

Scharff, Paul W. *Commentary on Rudolf Steiner's Agriculture Course: From the Paul W. Scharff Archive.* Great Barrington, MA: SteinerBooks, 2018.

Spindler, Hermann. *The Demeter Cookbook: Recipes Based on Biodynamic Ingredients.* Forest Row, UK: Temple Lodge, 2009.

Steiner, Rudolf. *Agriculture: An Introductory Reader.* Forest Row, UK: Rudolf Steiner Press, 2004.

———. *Bees.* Hudson, NY: Anthroposophic Press, 1998.

———. *Nutrition: Food, Health, and Spiritual Development.* Forest Row, UK: Rudolf Steiner Press, 2009.

———. *What Is Biodynamics? A Way to Heal and Revitalize the Earth.* Great Barrington, MA: SteinerBooks, 2014.

Strong, Devon. *A Lakota Approach to Biodynamics: Taking Life Seriously.* Great Barrington, MA: Lindisfarne Books, 2016.

Suchantke, Andreas. *Metamorphosis: Evolution in Action.* Ghent, NY: Adonis Press, 2011.

Thornton Smith, Richard. *Cosmos, Earth, and Nutrition: The Biodynamic Approach to Agriculture.* Forest Row, UK: Rudolf Steiner Press, 2010.

Thun, Maria. *The Biodynamic Year: Increasing Yield, Quality and Flavour: 100 Helpful Tips for the Gardener or Smallholder.* Forest Row, UK: Temple Lodge, 2010.

Thun, Matthias. *Biodynamic Beekeeping: A Sustainable Way to Keep Happy, Healthy Bees.* Edinburgh: Floris Books, 2020.

Wolff, Otto. *What Are We Really Eating? Practical Aspects of Nutrition from the Perspective of Spiritual Science.* Spencertown, NY: Mercury Press, 2023.

Wright, Hilary. *Biodynamic Gardening: For Health and Taste.* Edinburgh: Floris Books, 2009.

www.ingramcontent.com/pod-product-compliance
Lightning Source LLC
Chambersburg PA
CBHW052103230426

43671CB00011B/1910